D0014347

MANAGING
GEOGRAPHIC INFORMATION SYSTEM
PROJECTS

Spatial Information Systems

General Editors

P. H. T. BECKETT
M. F. GOODCHILD
P. A. BURROUGH
P. SWITZER

Managing
Geographic Information System
Projects

William E. Huxhold

Allan G. Levinsohn

New York   Oxford
OXFORD UNIVERSITY PRESS
1995

Oxford University Press

Oxford   New York
Athens   Auckland   Bangkok   Bombay
Calcutta   Cape Town   Dar es Salaam   Delhi
Florence   Hong Kong   Istanbul   Karachi
Kuala Lumpur   Madras   Madrid   Melbourne
Mexico City   Nairobi   Paris   Singapore
Taipei   Tokyo   Toronto

and associated companies in
Berlin   Ibadan

Copyright © 1995 by Oxford University Press, Inc.

Published by Oxford University Press, Inc.,

198 Madison Avenue, New York, New York 10016-4314

Oxford is a registered trademark of Oxford University Press

Library of Congress Cataloging-in-Publication Data
Huxhold, William E.
Managing geographic information system projects /
William E. Huxhold, Allan G. Levinsohn.
p.   cm.   —(Spatial information systems)
Includes bibliographical references and index.
ISBN 0-19-507869-1
1.Geographic information systems
I. Levinsohn, A. G.   II. Title.   III. Series.
G70.2.H89 1995 910'.285—dc20 93-43905

Page v constitutes a continuation of the copyright page.

7 9 8 6
Printed in the United States of America
on acid-free paper

Spatial Information Systems

The collation of data about the spatial distribution of significant properties of the earth's surface and of people, animals, and plants has long been an important part of the activities of organized societies. Until relatively recently, however, most of these data were kept in the form of paper documents and maps from which they could be read off easily, but only with difficulty could they be used to analyze the patterns of distribution of attributes over the earth's surface and the processes that had given rise to them. The developments in both computer technology and mathematical tools for spatial analysis that have taken place in the second half of the twentieth century have made many things possible, among them the ability to store, retrieve at will, singly or in combination, and to display data about all aspects of the earth's surface. As well as being able to handle existing data, the latest spatial information systems can just as easily handle fictional data and the results of simulation models, permitting scenarios of possible past or future situations to be modeled and explored. These abilities have created a revolution in the mapping sciences and in their uses in the practical day-to-day inventory, understanding, and management of our environment. Today, computerized spatial information systems are used in many branches of pure and applied science and in international government agencies. The applications range from the completely utilitarian, such as mapping the networks for telephones, electricity, and sewers, to the esoteric and futuristic, such as in modeling the possible future effects of climatic change.

The rapid growth in interest in spatial information systems and the accelerating rate at which they are being used has outstripped the normal rate of supply of trained scientists and technicians in the mapping and environmental sciences. Universities and polytechnics the world over are beginning to set up courses on spatial analysis and geographic information systems, not only to educate their own students but also to reeducate staff from government and business. Therefore, one aim of this series is to provide the basic source texts for the courses for third-year undergraduates, postgraduates, and practitioners, especially interdisciplinary texts covering basic principles that can be applied in many fields.

At present new knowledge about the theory and practice of modern spatial information systems is available only from conference proceedings and scientific articles published in a limited range of books and journals. The second aim of the series is to provide scholarly monographs, written by experts, in order to bring order and structure into this rapidly developing field. These monographs will gather new information from diverse sources and present it to researchers and practitioners so that it becomes widely understandable and available.

One of the most striking aspects of computerized spatial information systems is that, when they are used to bring data about many different kinds of spatial patterns together, the results rarely fail to surprise and delight. There is, of course, a danger that the user will be too easily impressed by sheer technology. It is the aim of this series to ensure that users of spatial information systems are not only impressed by technology, but are also really delighted because they have achieved a deeper understanding of the world around them.

P. H. T. Beckett (*editor*)
Department of Agricultural Science
University of Oxford

P. A. Burrough (*editor in chief*)
Professor of Geography
University of Utrecht

M. F. Goodchild (*editor*)
Co-Director, National Center for
Geographic Information and Analysis
University of California, Santa Barbara

P. Switzer (*editor*)
Professor of Statistics
Stanford University

Foreword

Successful implementation of geographic information systems is increasingly a function of management and institutional capabilities rather than that of technology. In the pages that follow, the reader will find a wealth of experience focusing on the context and content necessary to stage the introduction of GIS and its associated technologies into an organization.

The advice in this text is well tested. It springs from the practical experience of two professionals who have implemented GIS, successfully, over two decades. It is enriched by references to the works and views of many others. The presentation has equally been tested through interactions with both veterans and novices of GIS at The Banff Centre for Management in Banff, Canada.

Both managers and technical experts will gain from multiple readings and referrals to this work. Whether an organization is about to embark on an exercise to define and justify the applicability of GIS to its own organization, migrate an existing system to the current generation of practice and technology, or contribute to the activities of others, William Huxhold and Allan Levinsohn have provided a long-needed contribution.

John C. Antenucci
President
PlanGraphics Inc.
Frankfort, Ky.

Preface

There can be no doubt about it: GIS is more than a software package typical of word processing, spreadsheet, computer-aided design, and other popular computer products that one buys, installs, and learns how to use. Where, for example, does one obtain the map and attribute data to utilize the spatial query and spatial analysis capabilities of GIS? How does one ensure that the maps and data will be current and easy to obtain when they are needed? These are questions that users of those other software products do not ask, for they create the data needed in those products—all they have to do is to learn the commands and capabilities of the particular package they are using. They do not depend on others to create, update, and ensure that the data they need to solve their problems or help them understand issues related to locations on the earth are available when they are needed.

GIS is more than a problem-solving tool. Solutions to problems require knowledgeable people to analyze the situation, gather pertinent information, look at alternative solutions, and make a decision that, one hopes, is the most appropriate one. How do those people become aware of GIS capabilities and learn how to use them? Where do they get the pertinent information? Do they have to create the data, or do they have the data in an easily retrievable form? What if the GIS hardware or software does not work? Who is there to ensure that computer operations run as expected? These are questions that only the "GIS Elite" do not ask—they can do it themselves and do not depend on others to make GIS work for them. But at what cost, and where the heck are those people?

GIS, successfully implemented in an organization, is a technology that enables the "non–GIS Elite" to be more productive and more accurate in performing their work—without incurring excessive costs or adding additional complexities to their jobs. The successful GIS implementation institutionalizes the technology so that data, computer hardware and software, people, and procedures are available and dependable when they are needed. This book is addressed to organizations and individuals who wish to integrate GIS with their business activities and existing systems. It presents a framework and a set of techniques for planning, designing, and implementing a GIS to be shared by many users. It assumes that the reader is aware of what GIS is and what it can do. It assumes, also, that the reader wants to implement the technology in the most cost-efficient and appropriate manner. This book, then, is for those who want to know the "how" rather than the "what" with regard to GIS implementation.

Chapter 1, "The GIS Paradigm," sets the theme of the book: The successful implementation of GIS technology in an organization will address a variety of issues, including data management, information systems technology, and organizational strategy, tailored specifically to the particular environment of the organization. This chapter introduces the GIS paradigm as a foundation for using geographic information as a common base for managing geographic data, obtaining appropriate computer hardware and software to process those data, and providing the means to change existing responsibilities and procedures to use most effectively the technology for service delivery, management, and policy analysis within the organization. It introduces the concept of a data model: a conceptual representation of the information about the real world in which an organization operates, whether it is a government agency or a commercial enterprise.

Chapter 2, "Fundamentals of GIS Management," reviews the concepts of existing disciplines in information resources management and management science and practice and how they can be applied to GIS technology. It focuses on the decades of study of information technology and management science (before the development of GIS), as a precursor to understanding the use of GIS in an organization. It provides a list of nine "GIS Principles" developed by GIS and land records professionals that can be used as a framework for evaluating the technical details necessary for successful GIS implementation. It presents three case studies of the development of information systems (and geographic information systems) in local government: a $25 million failure of organization-wide information systems in the 1970s; a survey of existing (as of 1991) users of GIS technology to identify the most influential factors in the adoption and successful use of GIS technology; and the results of a study of nine agencies in Massachusetts and Vermont in 1990 that identified "organizational issues" as having profound influence on the success of implementation of GIS technology in local government. This chapter, then, draws upon preexisting understanding and current research on information systems and management practice in local government to formulate a foundation for successful GIS planning, design, and implementation.

Chapter 3, "Strategic Planning for GIS," presents a comprehensive, multiuser approach to planning the institutionalization of GIS. Its emphasis is on anticipating the short- and long-term issues that will be faced when a new technology is introduced. Central to this concept is that planning for GIS implementation in all aspects of an organization's business provides managers and workers with a new way of looking at the organizations's functions, creating opportunities to improve its operations—the paradigm shift. Strategic planning provides a framework for managing those changes and the interdependencies among the technology, the data, the people, and the organization. It discusses a situational analysis—understanding the organization in terms of organizational culture, management philosophy, and interpersonal relationships in evaluating how easy or difficult it will be to change the way work is done. It provides some models, based upon previous research, of the introduction and growth of computing technology in organizations and the factors that reduce its successful use. It also discusses the importance of a strategic vision for GIS—a general direction and long-term goals. This, it is argued, is necessary to build commitment and support within the organization, and it also helps align the direction of GIS with the strategic direction of the organization. Finally, to bring the vision in line with reality, the chapter addresses the feasibility study and the determination of the scope of the project—important information that decision-makers will need when they are asked to approve the GIS project. The chapter ends with a discussion of strategic planning for a multiparticipant GIS project when more than one agency contemplates a cooperative effort among agencies with common interests in geographic information.

Chapter 4, "Implementation Planning," focuses on a series of specific tasks necessary to analyze information and technology needs in order to design the system, prepare technical specifications, install hardware and software, convert data to digital form, and train staff. It describes the implementation-planning process as a framework for identifying the resources needed, a schedule of tasks required for completion of the project, the results expected when the system is operational, and the mechanisms that will be needed to manage the implementation process over a long-term time frame. This is a methodology chapter—one that addresses the major tasks to be completed to ensure successful implementation and use of the system. This chapter will guide the GIS project manager through the maze of activities that must take place once the decision-makers have approved the project.

Chapter 5, "Systems Design Methodology," provides a means to determine how the system will be used, what databases will need to be constructed, and what computer hardware and software will be required to store and process the data. It draws upon techniques used in the information systems field—information engineering—to ensure a comprehensive design that supports the critical functions of the organization. Its focus is on a hierarchy of analysis topics: how responsibilities (functions) of the organization determine what data are needed and how the data are to be processed to ensure the successful use of the system once it is implemented.

Chapter 6, "Implementation Management," takes the project from a design to an operational system that is successfully used by those for whom it is designed. It describes the processes of staffing and training, hardware and software procurement, the use of consultants and contractors, detailed design and implementation of the applications, and changes that will be required to the organizational structure leading up to the actual operation of the system. The emphasis of this chapter is on managing the implementation of the project— using the plans developed in the preceding chapters for what is to happen and providing the controls necessary to ensure that it does. The result is an operational system that is used to improve the efficiency and the effectiveness of the strategic activities of the organization.

Chapter 7, "Managing the System," describes the management issues that arise after the system is implemented and becomes operational. It addresses issues related to alternative possibilities for placing the GIS project in the organization; GIS personnel job descriptions, educational requirements, and management; work assignments and management of continued work projects; and budgeting and financial-reporting considerations. These are the important issues that need to be resolved when the effort changes from a project to a system.

This book uses its central theme—the GIS paradigm shift—to describe the activities and issues that are most likely to be encountered by the manager of a GIS project in preparing the organization for GIS implementation. This paradigm shift, the reorganization of the procedures and responsibilities related to the geographic information processing needs of the organization, is enabled by GIS technology to improve the efficiency and effectiveness of the workers, managers, and policy-makers of the organization.

Our interest in producing this text comes from many years of experience in managing projects, consulting with other managers of projects, delivering numerous seminars focusing on GIS management issues, and researching both GIS and IS (information systems) management issues. During those years we have been involved with many successful GIS projects, and we have been privy to a number of problems or failures in organizations that have begun GIS projects. In contrasting those successes and failures and in applying the valuable research that has been completed (and reported in this text), it has become clear that *how* an organization changes its procedures and responsibilities to enable GIS technology to move it toward a more productive and effective enterprise is much more complex and important than the technical aspects of *what* hardware and software are required. It is clear to us that the GIS industry needs this book.

Milwaukee, Wis.   W. E. H.
Canmore, Alberta   A. G. L.
July 1994

Acknowledgments

This book contains a compilation of knowledge and experience from many sources. Of particular value are the shared experiences from a series of GIS management seminars conducted at The Banff Centre for Management, Banff, Canada. Over 300 people from a wide variety of organizations have participated in those seminars since 1987. Their critical encouragement of our ideas and the sharing of their own experiences have contributed immensely to this work. Our thanks to all those who participated in the seminars and to The Banff Centre for providing the fabulous venue for those exchanges.

John Antenucci has been a major contributor to the Banff GIS seminars over the years. His practical insights drawn from his own experience in hundreds of projects are reflected in this book. He reviewed early versions of the manuscript and provided many helpful suggestions. We are grateful for his help in completing this work.

Stan Aronoff, friend and colleague, provided much needed advice and encouragement while we were completing the work. Thanks for your frank review of our early work.

Special appreciation goes to Dr. Peter Burrough, who introduced us and prompted us to write this work together.

Finally and most important, completion of this work would not have been possible without the dedicated efforts of Sandy Brown. Sandy merged our two writing styles, kept us organized, and managed the many details necessary in preparing a finished manuscript. We sincerely thank Sandy for her many contributions toward completion of this book.

Contents

1 THE GIS PARADIGM   **3**

The Four Elements of GIS in an Organizational Context   **5**
A Municipal Government Example   **14**
A Private-Sector Example   **18**
Applying the GIS Paradigm   **21**

2 FUNDAMENTALS OF GIS MANAGEMENT   **25**

Case Study 1. How Organization-Wide Information Systems
  Can Fail in Local Government: A Documented Example
  (USAC Experience)   **26**
Case Study 2. Correlates of GIS Adoption Success
  (Pinto and Onsrud Study)   **28**
Case Study 3. Organizational Issues in Successful GIS Implementation
  (Campbell Study)   **29**
A Need for Management Perspectives on GIS Implementation   **31**
Information Resources Management   **33**
Management Science and Practice   **36**
Conclusion   **40**

3 STRATEGIC PLANNING FOR GIS   **44**

GIS and the Organization   **47**
Strategic Planning: A Management Framework   **52**

Monitoring the Plan: Managing Interdependency   **56**
Selecting an Approach: Situational Analysis   **57**
Information Technology Factors   **59**
Applying the Results of the Situational Analysis   **65**
Strategic Vision: Building Commitment   **65**
Determining Feasibility: Establishing Project Scope   **67**
Strategic Planning for Multiparticipant Projects   **74**

4   IMPLEMENTATION PLANNING   87

The Implementation Process in Overview   **89**
The Implementation Planning Process   **89**
Creating an Implementation Management Framework:
   Assigning Roles and Responsibilities   **95**
Developing a Conceptual Overview   **102**
Managing Expectations, Establishing Priorities,
   and Establishing an Appropriate Sequence of Events   **107**
Preparing the Implementation Plan   **108**
Developing "Error-Free" Specifications   **112**
Implementing the Plan   **116**
Summary   **117**

5   SYSTEMS DESIGN METHODOLOGY   120

What Is Systems Design Methodology?   **121**
The Functional Approach to Systems Design   **127**
Applications Requirements Definition   **144**
Hardware and Software Requirements   **149**
Summary   **154**

6   IMPLEMENTATION MANAGEMENT   157

Managing the Transition to an Operational System   **160**
Staff and Training Needs   **164**
Technology and Services Procurement   **176**
Implementing the GIS Applications   **188**

7   MANAGING THE SYSTEM   203

Organizational Placement   **204**
Managing GIS Personnel   **207**

Managing the Work   **216**
Budgeting for GIS Operation and Maintenance   **223**
Summary   **226**

**APPENDIX: ADDITIONAL CASE STUDIES
AND EXAMPLES   229**

Government Mapping Agencies in Canada   **229**
Strategic Planning: Building a Municipal GIS,
    the City of Scarborough, Ontario   **230**
GIS Guidelines for Assessors   **231**
Examples of Procurement   **234**
Technology Selection Criteria
    by the Alberta Planning Board   **238**

Index   **241**

MANAGING
GEOGRAPHIC INFORMATION SYSTEM
PROJECTS

1

The GIS Paradigm

A geographic information system (GIS) is a collection of information technology, data, and procedures for collecting, storing, manipulating, analyzing, and presenting maps and descriptive information about features that can be represented on maps.

This definition covers a wide range of applications or uses of the technology, because a wide variety of organizations use maps and geographic data to support their activities. As a result, the use of GIS is becoming common in many different organizations involved with such diverse activities as surveying and mapping, forestry, utilities, transportation, insurance and finance, retail sales, and government. Many organizations involved in these and other activities are at various stages of investigating and implementing GIS technology to improve their operations. No comprehensive surveys have been made on the extent of GIS adoption among these sectors; however, experts will agree that there may be tens of thousands of organizations around the world that are involved with GIS to some extent.

Why GIS? Why have so many organizations purchased GIS technology? Presumably because they anticipate that GIS will provide new capabilities that will benefit the organization.[1] Examples of frequently stated benefits are

- Improved operational efficiency, including

  doing more with the same or fewer resources—for example, avoiding the addition of more clerical staff to deal with growing requirements for information products

  deriving greater benefit from staff activities—for example, making corporate information readily available to everyone, thereby facilitating greater delegation of authority

maintaining or reducing expenditures for data maintenance and related administrative tasks—for example, automating map updates

- Improved effectiveness, including

  an improved information flow to management and between departments—for example, organization-wide standardized databases, facilitating fewer custom reports and inaccessible information

  better utilization of staff—for example, staff who can devote more time to their key responsibilities and less to information gathering

  better decisions resulting from the review of more alternatives prior to committing to a course of action—for example, testing the impact on property values prior to developing a new commercial site in a residential area

In government, these benefits translate into an increased sense of well-being, including

- Increased wealth as a result of more effective expenditure of tax dollars and fewer administration costs, in turn providing funds to spend on other projects
- Better service to the public, directly through faster processing of permits and approvals, and indirectly through fewer service interruptions as a result of maintenance

For a company, these benefits translate into increased company performance, including

- Increased profits from more effective decisions and use of resources
- Making it easier to do business

These are lofty benefits to realize from the implementation of a computer system, and in fact it is well accepted that technology purchases alone will not produce these benefits. Thus it is the premise of this text that benefits such as those listed can be realized only by closely matching GIS with the operation of the organization. How to do that is the subject of this text.

Most early implementations of GIS were centered on map production and automation of the mapping process. Many other implementations were project oriented, wherein the system was used as a tool to solve an immediate problem or address a specific issue, and the data collected for the project were not used after the project was completed. In both cases, the GIS was used for a single purpose and the data were isolated from the main computerized databases of the organizations. In this single-purpose environment, the GIS has been separated from the daily activities of most of the staff in the organization. As a result, the benefits derived from its use have been limited, with few applications implemented.

Today, the nature of GIS implementation has changed as organizations have seen the need to integrate their GIS with other information systems, to move from project-oriented databases to continuously updated ("corporate-wide") databases, and to expand GIS use to decision-making and policy analysis rather than limiting its use to single-purpose support functions (such as mapping). Today's successful GIS implementations have integrated GIS into the daily operations of many of their organizational functions and are beginning to experience a variety of rewards and challenges in the process.

The Four Elements of GIS in an Organizational Context

As a first step in matching GIS technology to the business needs of an organization, a broader definition of GIS is needed, one that provides a comprehensive conceptual framework within which to discuss the institutionalization of GIS in an organization. In this text, GIS will be dealt with as the application of information technology, data management principles, and organization theory to the geographic information needs of an organization. These elements, focusing on geographic information, are discussed within the framework of a GIS paradigm: a set of geographic information processing concepts and principles that define a broad model of the real world within which an organization functions. Thus the four elements of GIS that this text uses in discussing the successful implementation of GIS consist of

- *The GIS paradigm.* A conceptual foundation for using geographic information that provides a common base of reference or focus for the other three elements
- *Data management principles.* The logical structuring and management of large databases that contain map and other data that can be related to the geography of interest to the organization
- *Technology.* The effective combination of various hardware and software components that enables the automation of numerous geographic data handling functions
- *Organizational setting.* A management environment that provides resources and enables changes to be made for incorporating GIS utilization throughout the organization

These four elements are tightly interconnected, as portrayed in Figure 1.1. Design and implementation decisions made in any one of these elements can directly affect one or all of the others. For example, the decision to include elevation data in the database can have an effect on how the data are to be stored and structured (as spot elevations or as contour lines). That affects the size of data storage devices and the requirements for manipulating the data (as a third dimension of coordinates or as an attribute). It also affects the cost of converting the elevation data from manual sources and the assignment of responsibilities for

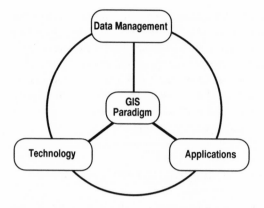

FIGURE 1.1 The elements of the GIS implementation framework.

updating the data when it changes in the real world. If these considerations are not addressed together as one issue, then the benefits associated with having elevation data available on the system are reduced or nonexistent altogether. Thus, to create a successful GIS implementation, it is important to manage all four elements together to ensure their alignment and compatibility throughout planning and implementation.

The GIS Paradigm

The *Oxford English Dictionary* defines a *paradigm* as a pattern or example. The term *paradigm* has taken on broad connotations in the discussion of changes in modern society.[2] In this book, the word *paradigm* is used to define a broad model or framework for understanding reality.

The GIS paradigm comprises the application of the fundamental geography[3] to organizing and using information. It uses space as the operational ground for solving practical problems. GIS, like geography, has as its focus the manipulation and analysis of data about features from the real world within a space–time framework. These concepts are not new: Only the application of information technology is new. In the past, the most common medium used to store and analyze geographic information was the paper map. Interpretation of the symbols representing various features and analysis of the relationships between features was performed by the human reading the map.

In the 1960s, concepts from map analysis were applied to various environmental and human settlement issues. One of the problems encountered was the integration of information from many maps with different themes. Although the themes shared common geographic space, the medium of the maps made integration of the information difficult. Early attempts at integration used transparent copies of various themes overlaid on a light table. The most widely recognized author of this technique was Ian McHarg (1969). Howard T. Fisher elaborated on these basic concepts and on the idea of applying the computer to such analysis and printed statistical values within a grid, on a standard line printer connected to a computer (Sheehan, 1979). The computer program was called SYMAP for SYnagraphic MAPping system. This event may have marked the birth of GIS technology. This work coincided with and was followed by that of many others (e.g., Steiner and Matt, 1972; Tomlinson et al., 1976; Fabos and Caswell, 1977; Teicholz and Berry, 1983).

From this early work, dealing with the application of geographic principles to the manipulation of geographic data, emerged a school of thought that produced formalized approaches to the combination of various themes of geographic information. These were later embedded in the software that now forms the basis of GIS technology as it is known today.

The concepts and principles that are fundamental to geography define the GIS paradigm, and are subsequently fundamental to successful implementation of a GIS. These are

- *Georeferencing*. The process of locating features within a model of the surface of the earth
- *Geocoding*. The process of attaching a geographic reference to nongeographic data
- *Topology*. The branch of mathematics that defines the relationships between features

These basic concepts, when properly implemented, create a model of the real world that can be manipulated and analyzed for the purposes of extracting information useful for making decisions about management actions in the real world.

As illustrated in Figure 1.1, the GIS paradigm defines how geographic data are represented and managed, determines the technology required to handle the data, and subsequently defines the actions required of the organization to achieve successful implementation. These elements are tightly linked: Decisions about any element affects the others. Design and implementation thus becomes an iterative process during which each element is investigated in relation to the others until a design that best aligns all the elements is produced.

GIS Data Management

Implementation of the paradigm begins by translating the user's perception of the real world into a model that can be implemented in a computer. Figure 1.2 illustrates the progression from perception to a physical implementation.

Different persons and organizations (or collections of people) perceive the real world differently. That is not to say that the world is different for each person, but that depending on education, responsibility, and the purpose of the organization where the person works, some parts of the world are of greater importance than other parts. This establishes a *perception structure* (mental

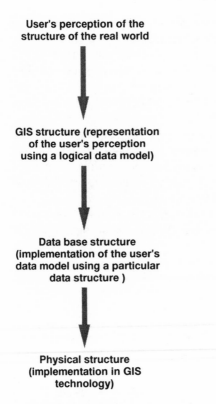

**User's perception of the
structure of the real world**

**GIS structure (representation
of the user's perception
using a logical data model)**

**Data base structure
(implementation of the user's
data model using a particular
data structure)**

**Physical structure
(implementation in GIS
technology)**

FIGURE 1.2 Implementation of the GIS paradigm in the computer. (Adapted from Burrough, 1986)

model) of the real world. That is, those features and events of significance to the individual or organization, often differ due to management responsibility. To enable the use of computers, the perception structure must be translated into a format that can be understood and worked upon by the computer.

To accomplish this, the perception structure is first organized into a conceptual or *logical data model* that represents the application of the GIS paradigm to the user's perception structure. The data model consists of two main parts: *spatial entities,* which are computer representations of features from the perception structure (e.g., a lot, a forest stand, a section of pipe, an oil well), and descriptive data about the entities, commonly referred to as *attributes.* The data model describes the entities and attributes, their relationship to one another, how they are used by people, the processes that are used to manage them, and how the data are moved about during use.

A common data model does not equate to one database. However, implementation of the data model does require adoption of some standards. The standards deal with how data are defined, stored, and moved between systems or applications. The standards can be designed to provide the optimum balance between sharing and individual use by defining the minimum requirements for data sharing. This concept can be thought of as a "common data platform" upon which individual users can add their specific data as required. The common data platform is derived from the understanding of what is common and what is unique among data users. The data model becomes the means to determine the best balance between commonality and specificity.

Once defined, the logical data model is translated into a database structure that can be implemented on a computer. There are many database structures used to represent geographic data, depending on the requirements of the data and its subsequent use. These are not dealt with in this book, nor are the physical structures implemented in the GIS technology. For information about these topics the reader is referred to the references in the note at the end of the chapter.[4]

Common to all geographic data structures is the concept that the map becomes a set of spatial entities or layers of elements that, when combined in an appropriate manner, can create a map. There is no map as such. The map is a virtual map and consists of logical elements stored as computer representations and their relationships. The purpose is to imbed in the computer the ability to assemble and analyze the entities and attributes in a manner that resembles the mental processes performed by a human reading a map. Once established, the variety of descriptive information can be stored about each entity, and the information can be retrieved in a wide variety of ways.

Figure 1.3 shows a simple conceptual geographic data model. Note that all the spatial data share the same positional base and that common georeferencing and standard definition of data facilitates integration of a wide variety of spatial and attribute data. A user of this information could begin by identifying a specific location and determining data relative to that location, or the user could select a topic such as buildings and determine the locations that correspond to various types of buildings. Figure 1.3 is a conceptual illustration, not a representation, of how the data model would be implemented. The reader is again referred to other sources for information about data structures and their implementation.

An important aspect about data describing the real world is that the real world is dynamic. Changes in the real world need to be reflected in the data for it to remain useful for decision-making. Thus a means of updating the data and keeping it current with real world events is required. A *data management strategy* provides a set of guidelines for structuring the collection, management, and storage of data. The guidelines are implemented within a management framework that defines how the processes used to collect, maintain, and manage the data will be organized and performed.

The data management strategy affects the allocation of staff and funds and therefore impacts the organization's management structure. The data management strategy will also affect the type and distribution of technology, which, in turn, affects the operation of the organization. For example, if production rates from gas wells are currently recorded on paper data collection forms and sent

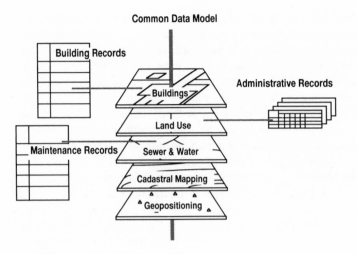

Requirements for a common GIS data
model:
• Shared geopositioning
• Standard data definitions
• Explicit entity relationships
• Planned data distribution
• Standards for data communication
• Data maintenance processes

FIGURE 1.3 Conceptual illustration of a GIS data model.

to a data center for input to the computer, and the new data management
strategy requires them to be input at the well site from remote terminals over
communications lines, the new strategy impacts both the work of the organization and the type and location of technology. Data center staff will need to be
reassigned, and field staff will need to be trained in the use of the new
technology. The selected GIS technology will need to include an appropriate
device for field data recording and a network for data communications.

The GIS paradigm is made tangible through its implementation in data. A
data model defines the entities, attributes, and relationships among them that are
of significance to an organization. The data model is subsequently implemented
within a selected technology.

GIS Technology

Modern information technology makes implementation of the GIS paradigm
practical. The rise in GIS over the past 3 decades has been paralleled by the
extraordinary gains in computer performance. Since 1988, price and performance have been improving at a 6 percent compound-growth rate every month,
leading to a doubling of performance every year (Goodman and Sproull, 1990,

p. 235). Since the early work at the Harvard University Laboratory for Computer Graphics and Spatial Analysis in the late 1970s, software developers have capitalized on hardware gains, and many spatial analysis algorithms that initially ran as batch operations overnight can now be processed in minutes or seconds as on-line operations. The implementation of automated mapping had immediate appeal for many organizations, and many were quick to acquire the new technology. However, many early implementors discovered that high-performance technology alone was not enough for successful GIS implementation.

In this book GIS technology will be dealt with as the enabling portion of the GIS paradigm. Without the technology, implementation is not practical. Technology that is well aligned with the data and organizational requirements will enable successful systems to be built. However, too much emphasis on the technology, especially in the early stages of GIS implementation, will lead to expensive setbacks. Technology selection should be the result of careful analysis of data management requirements and organizational impact. The process of determining and implementing the best technology is a primary topic of this book, and it will be dealt with repeatedly in subsequent chapters.

To begin with: What is GIS technology? Few would disagree that computer-assisted mapping is an integral part. The need to store and manage attributes provides the basis for including database management software in GIS technology. Modern digital instruments to collect geographic coordinates might also be included (e.g., global positioning systems [GPS][5] technology). Some GIS applications now include video images and sound, and on that basis multimedia technology might also be included. If an application, such as official notification, includes automatic generation of a form letter to affected rate payers, then perhaps word processing or text management might also be included in a GIS application. Table 1.1 shows a list of commercially available technology that might be used in a GIS application.

The purpose of this discussion is to suggest that, for various reasons, the optimum technology to satisfy a particular set of needs may go beyond digital mapping and database management. In this book, GIS technology implementation will emphasize selecting a combination of information technology that best enables automation of all or part of a selected business function. As will become evident in later discussions, this perspective on GIS technology significantly expands available alternatives for data conversion and management.

Organizational Strategy

In order to benefit from the full potential of GIS technology, it must first be understood that GIS is not merely another software product purchased by an organization to add to its arsenal of word processing, spreadsheet, and other packages for an organization's employees to learn and use. While GIS functionality and its ability to provide simple user access to geographic data are very important to the success of a GIS project, there are broader, more basic,

Table 1.1   Candidate Information Technology for GIS Implementation

Technology	GIS Application
CAD (computer assisted drafting)	Digital mapping
DBMS (database management system)	Manage and retrieve land-related records
Geoprocessing (spatial analysis) using	Various land and resource analysis models such as operators as overlay, proximity
Remote sensing and image analysis	Display and analysis of geographic images linked with other geographic data
GPS (global positioning system)	On-site collection of geographic coordinates
Multimedia (e.g., sound, video)	Video image of pavement condition referenced to road segments
SCADA (supervisory control and data acquisition)	Water or gas valve control from geographic database of valve locations
Document imaging	Store and retrieve card images of maintenance records; store images of building floors
Text processing	Automatically generate letters of notification to property owners determined from a spatial query
Network communications and electronic data interchange (EDI)	Link databases and users in different locations such as separate parts of an organization or different organizations

and long-term issues that must be addressed in implementing GIS technology in an organization. The issues that must be addressed concern how the technology can be incorporated into the working environment of the organization: how it fits into its broader information systems (IS) environment; how it fits into the organization structure; and how it will help the organization manage information about the real world as required to deliver products or services. These issues rise above the technology and the practical issues of the cost and time needed for implementation. They are even more important than the education and training needed for the users to operate the system fully.

Organizational structure enables the organization to fulfill its mission or mandate. Public agencies educate people, fight crime, put out fires, combat health problems, provide recreational facilities, and deliver other similar public services that relate to people, property, facilities, and economic conditions in the real world. Private companies make products, sell them to customers, provide services, and, it is hoped, make profits. In this context, a GIS is a specialized part of the organization's information systems environment that supports the organization in performing some functions in response to changes in the real world.

Different organizations operate on different parts of the real world, and respond to different perception structures. These different requirements translate into different organizational structures and cultures and thus into different information systems environments. The perception structure of an international manufacturing organization is quite different from that of a local government organization, which, in turn, is different from that of a national government agency. The international manufacturing organization reacts to market forces, new manufacturing techniques, local and worldwide economic conditions, and other private sector stimuli. Local governments, on the other hand, address the changing needs of citizens and the physical features within its jurisdiction, such as the condition of sewer mains, fire response reports, and property tax rates. The focus of a national government is protecting the environment, supporting transportation systems, and national defense, to name a few. The organizational structures and the information systems of these organizations are therefore necessarily different. Similarly, a GIS to support the needs of each organization would also be different.

Successful GIS implementation, in part, depends upon understanding the real world in which the organization operates—how changes to real world features affect the information needed by the organization and how information systems are used to support the functions that deal with such changes. Developing an understanding of the perception structure of each organization enables the parameters determining organizational structure, information systems, and subsequently GIS, to be defined.

During the process of GIS design and implementation, the perception structure of an organization is translated into a series of successively more detailed specifications. The format of the process and the design of the GIS should mirror the requirements of the organization as defined by its perception structure. To understand how data should be structured for the efficient and effective operation of the organization, it is helpful for one first to develop a conceptual model of the entities and conditions in the real world environment of the organization, and second develop a model of the attributes of those entities and conditions needed in its operation. This process of conceptualization should be undertaken without regard to existing organizational structures because the real world entities and conditions exist regardless of how the organization is structured. They are not affected by the structure of the organization; they are affected only by the results of the operation of the organization.

For example, whether a municipality conducts building permit inspections within its Public Works department or in a completely separate building inspection department, local laws and ordinances normally require buildings to be inspected when certain changes occur to them. How well the inspections are conducted—how efficient and effective is the inspection process—depends, to a great degree, on how the municipality is organized. But the fact that inspections must be made does not depend on how the municipality is organized. The conditions causing changes to the buildings (such as remodeling, damage, change in ownership, change in use) exist in the real world regardless of the organizational structure of the municipality. In this case, it is important first to identify what real world changes to buildings create a need for inspections, and second to identify what information is needed about the buildings and their changes to aid the municipality in its inspection responsibilities. This should be identified without letting existing organizational structures and practices restrict, constrain, or influence the identification of what will eventually be information and information processing needs.

The construction of a data model, then, for GIS implementation and use begins with the creation of a model of the organization's environment—how the real world entities and conditions within the jurisdiction of the organization behave and interact with each other over time. This allows the definition of the information needed by the organization to be easier to identify, to be more comprehensive, and to become less influenced by inefficient or inaccurate practices currently within existing organizational structures. The result is a data model that, when used in the design, development, and operation of the GIS, will withstand any future organizational and procedural changes associated with the form of the organization.

A Municipal Government Example

The preceding discussion has established a general conceptual basis for establishing the parameters of a GIS. In the following example, this conceptual framework is applied to the perception structure of a municipality.

The physical components of a municipality are

- The *geographic coverage* of its jurisdiction: political boundaries, natural features
- The *people* within that area: residents, business people, visitors
- The *buildings and facilities* that the people use: buildings, streets, utilities
- The *business activities* that influence the lives of the people and their physical environment: jobs, goods, services
- The *land* within the geographic coverage that supports the physical objects and activities of business and people

These are the entities and activities of the environment in which the municipality operates. They are, in fact, the reason that the municipality exists: to bring order to the physical, social, and economic environment of its geographic coverage to provide for a viable community.

Now consider what changes occur to these entities that are of interest to the municipal government in meeting its objectives (Figure 1.4).

- *People.* Consider what changes to people occur over time within the municipality:

People move in; people move out.
People are born and die.
People are educated, get jobs.
People buy land, goods.
People become sick, jobless, married.
People commit crimes, occupy buildings, own pets.

- *Buildings and facilities.* Consider the changes that occur to the physical objects in a municipality that people use. Consider the different states that they are in over a long period of time:

  Buildings are built, occupied, deteriorated, demolished.
  Streets are constructed, used, repaired, replaced.
  Trees are planted, maintained, damaged, removed.
  Traffic lights are installed, repaired, replaced.
  Sewers are designed, laid, maintained, replaced.

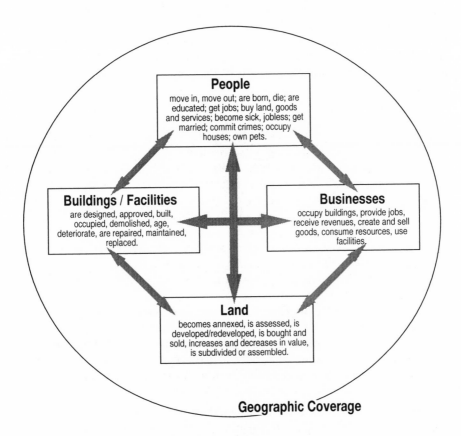

FIGURE 1.4 Municipal government perception structure.

- *Business activities*. Look at the changes occurring within the jurisdiction that affect the business activities within the municipality:

  Businesses are created, occupy buildings, operate, disband.
  People get jobs, spend money.
  Goods are manufactured, sold, bought, consumed.

- *Land*. Consider the changes that occur to the land within a municipality over time:

  Land becomes annexed, detached, zoned, rezoned.
  Areas are developed and redeveloped.
  People and businesses buy and sell properties.
  Values increase and decrease.
  Parcels are subdivided and assembled.

Now consider all the interactions among these elements within the geographic coverage. This is the environment in which the municipality operates. This is a perception structure of the municipality's real world.

In local government, a municipal organization has certain responsibilities for the environment as described so far. A municipality performs services, regulates changes, enacts and enforces laws, and constantly seeks improvements in the physical, social, and economic viability of the community. It also collects taxes and fees to raise the revenue needed to operate. This is the business of government.

In the process of conducting its business, a municipality also collects or otherwise obtains information about the people, buildings, facilities, land, and business activities within its geographic coverage. This information is used in many different ways to provide services, manage public resources, and establish plans and policies that are intended to improve conditions of the city. Information about reported crimes, for example, is used not only to dispatch police and make an arrest that can eventually result in a conviction, but it is also used to

- Ensure that enough police are available in the right areas at the right times to respond to other requests for service
- Aid in the investigation of other crimes that may have been committed by the same individual
- Evaluate the effectiveness of police services (response time, arrest rate, conviction rate)
- Determine the need for changes in laws and penalties for breaking them
- Evaluate the impact of changes in policies and procedures, new programs (street lights, block watches) or even police training courses

Similarly, information collected about buildings is used not only to determine their value for establishing fair and equitable taxes, but also to

- Determine how many and what types of property assessors are needed to perform what activity
- Identify who owns which properties for sending building code violation notices, notices of zoning changes, and notices of public meetings
- Determine the appropriate fees for water and sewer usage, fire inspections, future public facilities such as schools, infrastructure, and recreational facilities
- Determine current housing stock availability in the process of planning future growth incentives
- Assess the impact or need for new laws and ordinances such as the installation or fire alarms, facilities for the physically challenged, and building codes

The product of identifying and organizing the information needed by local government to deal with the changes to the components of its jurisdiction is a data model. The data model provides a high-level definition of the range of information about the people, property, and conditions of the municipality that are needed for the government to operate and that will eventually be stored and processed in a GIS. The content of the data model depends upon the perception structure of the organization and what activities the organization performs that affect the objects and events that comprise the perception structure.

The activities of a municipality provide a rational limit, or range, on the extent of information needed in a GIS. These functions define which people, property, and conditions are important in collecting and maintaining information for a GIS. Notice, at this point in the data modeling process, that the existing organizational structure of the agency is unimportant—only the functional responsibilities are important. For example, it is not important yet to identify which office is responsible for crime-related activities or for building-related activities, but it is important to identify that both functions will need the same information in many activities (such as building address or owner name). In the examples given the following functions provide a framework for defining the information needed in the data model about crime and buildings:

- Dispatching police to the scene of a crime
- Obtaining sufficient police resources
- Solving crimes
- Training police personnel
- Enacting laws
- Implementing programs to reduce crime
- Assessing property
- Collecting taxes
- Inspecting properties
- Establishing zoning regulations
- Planning for public facilities

The data model provides for the attributes of the entities that are subjects of the functions of the organization. For example, the attribute "address" is needed in the two examples for dispatching police, solving crimes, assessing property, inspecting properties, and, possibly, establishing zoning regulations. "Type of crime" may be needed to obtain sufficient police resources, solve crimes, train personnel, enact laws, and implement programs to reduce crime.

Collecting and maintaining information about people, property, and conditions of the municipality that are not subjects of its governmental functions is an inefficient use of the organization's resources. Is it important for a municipal government to know how many of what type of trees are on the properties of its residents? Maybe not. Maybe it is important for the municipal government to have information only on the trees that are on public land, because it is responsible for maintaining those trees. It is not responsible for maintaining trees on private land and so would not require information on them. When the functions of the organization are identified and related to the model of the real world of its jurisdiction, the framework for defining the data model begins to take shape.

A Private-Sector Example

In the private sector, the real world of a commercial enterprise is much different. Although businesses usually operate within a geographic area and interact with physical components (entities) that are essential for their successful operation, these entities and the changes they experience over time in the real world are different from those in municipal government. For example, a business may not be affected by the political jurisdiction (corporate boundaries) of municipalities within its market area; the potential market of the business may not include all people within its geographic coverage; it may see an opportunity to expand (or need to reduce) its market area; it must be aware of its competition and the changing needs of its customers; it must ensure that the appropriate resources are available to create and move its product to the customer. The difference between the public sector and the private sector is not so much profit versus nonprofit—the difference is the reason they exist and the environment (their real world) in which they operate. Thus their organization structures and the supporting information systems are different. This results in different data models between the two types of organizations.

Consider, for example, the typical business operation. The physical components of its real world might include the following entities:

- The *geographic coverage* of its operation: market area (region, state, country), sales territories
- The *market* for its goods and services: people, organizations
- The *employees* it uses to conduct business: sales staff, production staff, support staff, contractors (brokers, partners)

- The *facilities* it needs to create and move the product to the customer: stores, offices, production facilities (buildings, equipment), distribution facilities (warehouses, vehicles)

These are the entities that are important to the operation of a business because it could not exist without a market for its product, customers to purchase the product, and employees and facilities to get the product to the customers. This real world of the business concern is much different from the real world of a municipality.

Consider now the changes taking place to these entities that are important to the business concern (Figure 1.5).

- *Market.* The people and organizations that could become customers experience changes that can affect whether or not they do become customers:

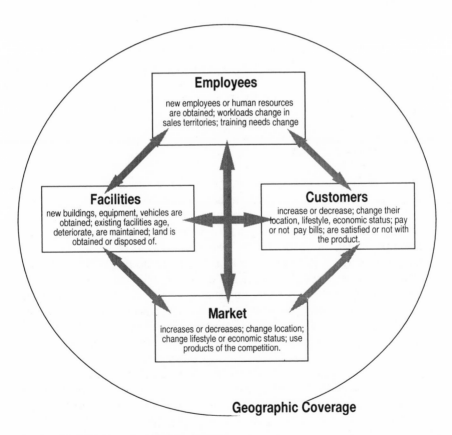

Employees
new employees or human resources are obtained; workloads change in sales territories; training needs change

Facilities
new buildings, equipment, vehicles are obtained; existing facilities age, deteriorate, are maintained; land is obtained or disposed of.

Customers
increase or decrease; change their location, lifestyle, economic status; pay or not pay bills; are satisfied or not with the product.

Market
increases or decreases; change location; change lifestyle or economic status; use products of the competition.

Geographic Coverage

FIGURE 1.5 Private-sector perception structure.

They increase or decrease in numbers.
They change their location.
They change their lifestyle or economic status.
They use products of the competition.

- *Customers.* The existing customers or clients of a business experience changes that may or may not be similar to the changes taking place in the larger market:

    They increase or decrease in numbers.
    They change their location, lifestyle, economic status.
    They pay their bills or fail to pay their bills.
    They are satisfied or dissatisfied with the product.

- *Employees.* The employees of a business have a major impact on its successful operation and thus changes to these human resources affect the real world of the organization:

    New employees or human resources are obtained.
    Workloads change in sales territories.
    Training needs change.

- *Facilities.* The buildings, equipment, vehicles, and other physical resources used in the creation and distribution of the product experience changes that must be managed by the organizations:

    New buildings, equipment, vehicles are obtained.
    Existing facilities age, deteriorate, are maintained.
    Land is obtained or disposed of.

Of course, external influences play a large role in the success of a commercial activity. The changes that take place in the commercial-sector environment that affect successful operations include government regulations and incentives; taxes and the economy, new technology and transportation facilities, and political relationships among various levels of government.

By considering all of the interactions among these entities and their activities (including external influences) in terms of a particular business concern, a conceptual model of the real world environment emerges, just as it did in the example of the municipal government. This model, however, is quite different from the municipal model and so its perception structure is different. This results in a different organization structure and environment, which, in turn, creates a different information system design to support it.

In the private sector, the business organizes itself in a manner that best addresses its mission in its real world. Typically, it will have a marketing or sales function to expand and maintain an income stream; it will have a production function to create and improve its product line; it will have a distribution mechanism (retail outlets, transportation facilities) to ensure that the product can be moved to the customer; and it will have internal functions (personnel, accounting, research, design, maintenance) to support the people and facilities involved in producing, selling, and distributing the product.

As in the municipal government example, the business collects or obtains a variety of information about the entities and their changes in the real world during the operation of its functions. It obtains information about its customers (who, where, when, sales volume, product types, billing and payment status), information about its market segment (demographics, buying patterns, econometrics), information about its product (sales by territory, sales person, or store), information about its facilities (fuel and maintenance), and other information that is necessary for the function to be effective. This information is used in many ways in the numerous activities of the organization to support its mission. This information is the data model of the business.

In both the municipal government and the commercial organization examples, the organizational strategy described involved the identification of the entities and their changes that are important in the real world of the organization. This set the stage for defining what functions and activities are important for the organization to complete its mission within its real world and then identifying what information is necessary in those functions and activities to be effective. The identification of the information about the entities important to any particular organization becomes the data model of the organization: a record of the entities and the attributes of those entities that are important in performing its functions.

Applying the GIS Paradigm

When an organization introduces a new information technology such as a GIS to an organization—one that has the potential to be used by many different functions in the organization—a data model helps to ensure that the design of the system can support the many diverse needs of the organization. The organization, however, must be flexible enough to change its structure and data management responsibilities in order to take advantage of the new technology and the new flow of information that will most likely result. The GIS paradigm is a concept that helps identify these changes and achieve balance among the technology, the data, and the organization structure when GIS technology is implemented. It identifies better ways to collect and manage geographic data by using new information technology and by changing existing activities to become more effective.

The process and techniques used to analyze, define, and operationalize the data, technology, and organizational structure for successful GIS implementation is the subject of this book. Figure 1.6 presents an overview of the topics and their organization in this book.

GIS implementation in the broad context presented in this chapter requires consideration of subjects that have been well developed in other fields of study.

	GIS Concepts & Principles	Organizational Theory	Management Science and Practice	Management Information Systems	Information Technology Planning
Chapter 1: The GIS Paradigm	GIS as an Abstract Model of the Real World	Benefits to the Organization from GIS			Subject Flow
Chapter 2: GIS Development	GIS Principles	Organizatonal Change & GIS Implementation	Management Practice & GIS Implementation	GIS as a Management Information System	Information Technology Paradigms Applied to GIS
Chapter 3: Strategic Planning	GIS in the Organization	Building Vision & Commitment	Selecting an Approach to GIS Implementation	Aligning GIS with MIS Planning	Developing a GIS Technology Strategic Plan
Chapter 4: Implementation Planning	Criteria for GIS Applications	Organizational Structures for GIS Implementation	Building Project Teams	Quality Systems & Specifications	The Planning Process
Chapter 5: Managing the Design Process	Defining a Geographic Data Management Process	Change Management & Managing Expectations	Project Management Tools & Techniques	Defining Needs, Opportunities & Feasibility	Data & Functional Analysis for Application Design
Chapter 6: Implementation Management	Implementing the GIS Paradigm	Maintaining Commitment/ Training & Education	Project Management	Purchasing GIS Products & Services	Application Construction
Chapter 7: System Operation & Management	Fulfilling the Vision	Staff Development & Career Planning	System Operation Administration & Maintenance	Managing Service Delivery	System Evaluation, Migration Planning & Enhancement

Process Flow

FIGURE 1.6 Overview of the topics in this book.

An overview of each subject is presented in Chapter 2 as the basis for discussion of related topics in later chapters and to provide a framework for organizing the subjects into a process for GIS implementation.

In the ensuing chapters, the GIS paradigm is applied to GIS implementation. In Chapter 3, an approach for establishing the role of GIS in the organization is presented. The result is development of a multiyear strategic plan. The strategic plan is translated into an implementation plan in Chapter 4. Chapter 5 then deals with the details of determining the organization's needs for GIS. In Chapter 6, the framework and techniques established in Chapters 4 and 5 are applied to implementation management. Finally, in Chapter 7 management of the operational GIS is dealt with. Each of these is built upon the concepts presented in this chapter and in Chapter 2.

NOTES

1. In this discussion, *organization* is used as a general term describing any group of people and organizational units sharing a common purpose; it could be a company, government agency, or collection of agencies.

2. Tapscott and Caston (1992) describe four paradigm shifts influencing business today:

- A change in the world economic and political order—a shift to a more open, volatile, and multipolar world
- A change in the business environment and marketplace—markets and national economies are being transformed
- A change in the nature of organizations—a shift to a more dynamic enterprise, enabled by information to succeed in the new environment; flatter and team oriented; based on commitment; streamlined, open and networked
- A change in information technology—a shift to more open, usercentered, and networked computing

3. Strahler (1969) describes geography as the study and unification of a number of earth sciences that give general insight into the nature of man's environment. Because data of the earth sciences are often best presented by maps, and perhaps many are impossible to describe without maps, the science of maps, *cartography,* is an essential ingredient. Muller (undated) adds that geography focuses on the analysis and manipulation of entities within a space–time framework. Typical situations where GIS are used, such as urban and regional development planning, monitoring of changes in land use and land cover, conservation and exploitation of natural resources, and assessment of environmental changes (deterioration of soils, pollution, flood hazards), have belonged to the realm of geographic interests for a long time. Coordinate transformations, polygon overlays, classifications, generalizations, spatial aggregation and districting, proximal and contiguity analysis, optimal path finding are data handling procedures familiar both to GIS and geography.

4. For references on database structure and physical structure in GIS technology, the reader is directed to Aronoff (1989), Burrough (1986), and Maguire et al. (1991).

5. Global Positioning System (GPS) operates through a constellation of satellites originally developed by the United States Department of Defense as a navigational aid that recently has become available for geodetic control surveying, location of features, and "on-site" digitizing. The satellites transmit signals that can be decoded by specially designed receivers to determine positions of varying accuracy (depending on the device and the existence of ground stations). These measurements can be made at a fraction of the cost of traditional first-order surveying techniques. Eventually, the existing satellite constellation GPS network will be replaced by an 18-satellite constellation that will allow 24-hour, all-weather operational capacity in both navigation and relative positioning.

REFERENCES

Aronoff, S. (1989). *Geographic Information Systems: A Management Perspective.* WDL Publications, Ottawa.

Burrough, P. A. (1986). *Principles of Geographical Information Systems for Land Resources Assessment*. Oxford University Press, Oxford.

Fabos, J. G., and S. J. Caswell (1977). Composite landscape assessment: Metropolitan landscape planning model MET-LAND. Massachusetts Agricultural Experimental Station, University of Massachusetts at Amherst Research Bulletin no. 637.

Goodman, Paul S., and Lee S. Sproull (1990). *Technology and Organizations*. Jossey-Bass, San Francisco.

Maguire, D., M. Goodchild, and D. Rhind (1991). *Geographical Information Systems: Principles and Applications*. Wiley, New York.

McHarg, I. L. (1969). *Design with Nature*. Doubleday/Natural History Press, New York.

Muller, J. C. (undated). Geographic information systems: A unifying force for geography. Paper, University of Alberta.

Sheehan, D. E. (1979). A discussion of the SYMAP program. Harvard Library of Computer Graphics, mapping collection. Vol. 2: Mapping software and cartographic databases, pp. 167–79.

Steiner, D., and O. F. Matt (1972). Computer program for the production of shaded cloropleth and isarithmic maps on a line printer. User's manual. Waterloo, Ont.

Strahler, Arthur N. (1969). *Physical Geography*. 3rd ed. Wiley, New York.

Tapscott, Don, and Art Caston (1992). *Paradigm Shift: The New Promise of Information Technology*. McGraw-Hill, New York.

Teicholz, E., and B. J. L. Berry (1983). *Computer Graphics and Environmental Planning*. Prentice-Hall, Englewood Cliffs, N.J.

Tomlinson, R. F., H. W. Calkins, and D. F. Marble (1976). *Computer Handling of Geographic Data*. UNESCO, Geneva.

2

Fundamentals of GIS Management

In Chapter 1 the GIS paradigm was introduced, establishing a broad context for GIS implementation. The premise presented was that successful GIS implementation depends as much or more on successfully dealing with planning, design, and institutional issues as it does on GIS technology.

In this chapter that premise will be expanded, arguing that a successful GIS is built, not bought, that a systematic approach to implementation will yield better results, and that significant commitments are required from all levels in the organization or organizations undertaking GIS implementation.

Although GIS is a relatively new information technology application, many of the supporting concepts are not, as discussed in Chapter 1. Similarly, much can be learned about introducing new technology into organizations and implementation management from areas of study not directly related to GIS. Information technology has been in use for approximately 30 years, and a large body of knowledge and experience has been accumulated by the information technology community. Similarly, management science and other related disciplines have been studying organizational change and the introduction of new technology for decades. This chapter will explore what can be learned about successful GIS implementation from these experiences.

To begin this discussion, three case studies are presented. These three studies are all from local and state government; however, it is reasonable to extrapolate these results to other organizations. The three studies explore how medium to large organizations have dealt with information systems development, and summarize their experiences.

Case Study 1. How Organization-Wide Information Systems Can Fail in Local Government: A Documented Example (USAC Experience)

In 1970, six medium-size cities in the United States were awarded $20 million by the federal government to develop integrated municipal information systems (IMPS) for their agencies. The Urban Information Systems Inter-Agency Committee (USAC) was established to fund and guide the effort in applying integrated information systems concepts to urban management in anticipation of improving local government planning, management, and operations and then transferring these applications to many other local governments across the nation. The underlying philosophy of the USAC Program is best documented in "A Requiem for USAC":

> The USAC architects further believed that the most promising tool for achieving this improvement was increased use of computers and automated information systems: information systems could automate routine information processing tasks, thereby improving productivity in the information-dependent field of local government management; they could be used in analysis and redesign of municipal goals and activities; they could integrate data and data processing, creating a dynamic data base for use in planning, management, and integration of local operations. Ultimately, the designers of USAC felt that the use of computers could result in development of new sociotechnical systems offering increased local government efficiency and improved service delivery. (Kraemer and King, 1979, p. 315)

The technical concepts of the IMPS plan reflected the state-of-the-art in information systems design at the time, defining for the first time a complete data model and information flow requirements of an entire municipal government, thus integrating data processing across an entire organization. These concepts included

- A total municipal information system based on four major business areas: public safety, public finance, physical and economic development, and human resources
- A single database that could integrate functions both vertically among government functions and horizontally around the objects of people, property, money, and personnel
- Data based upon the operational data files of municipal agencies because planning and management data are considered to be byproducts of operational data
- Top–down design, from conceptualization, analysis, design, development, implementation, and evaluation, with prototypes implemented incrementally
- Complete documentation at all phases to facilitate transfer to other municipalities
- Flexible design to facilitate transfer to other municipalities

- A consortium approach to project management, with the city providing project leadership, a consulting firm providing technical expertise and training, and a university providing technical advice and evaluation

Thus, with a significant amount of funding, clearly defined goals, and state-of-the-art information systems design concepts, the six cities embarked upon designing Integrated Municipal Information Systems.

After 7 years and $26 million, the program failed and the USAC committee disbanded after a panel of experts found that "virtually no total systems had been transferred from the USAC cities to other local governments" (Kraemer and King, 1979, p. 323). In their requiem, Kraemer and King cite the following problems with the USAC program:

1. The operational meaning of the USAC concept of integration was more complex than anticipated, disrupting existing procedure and practice by imposing new standardized data-sharing regimen.
2. The presence of data for planning and management is only one criterion for its successful use—analytical staff are needed to make extensive and imaginative use of the data and mechanisms are needed to keep track of the data available, ensure the data are current, and massage the data to produce information.
3. There was no unifying or directing force, such as a single client; so the activities of the project could not be centered on a systematic and stable environment.
4. The top-down approach proved too cumbersome and costly, precluding the opportunity to implement some applications early in order to demonstrate benefits.
5. The top-down approach was abandoned when the mass of detail became overwhelming: Designers began to develop only those applications that they intuitively felt were needed when deadlines approached.

In summary, the USAC program demonstrated that a "supply–push" strategy that assumes that there is an existing universal demand for integrated municipal information systems was wrong. It demonstrated the difficulty of encouraging organization-wide information systems simply through providing funding and state-of-the-art technology. As Kraemer and King put it: "Thus, a specific package of information management technology is not likely to be in universal demand" (p. 347).

Huxhold (1993) recommends that GIS proponents need not despair from the USAC failure, but rather learn from it. A number of lessons learned from that experience can assist GIS proponents in developing successful organization-wide geographic information systems:

1. An organization is not ready to adopt GIS technology because of its

potential alone. It is ready to adopt the technology when it has identified a problem in its own organization and sees GIS as a solution.

2. A common goal that reflects the individual goals of each participant provides a basis for shared GIS development. These goals must be aligned with the overall goals of the organization.

3. The scope of the GIS project must be clearly understood by all participants, including an understanding of the potential for changes in organizational structure and procedures.

The preceding study relates to systems that developed in the 1970s involving a complex set of integrated functions. What about GIS technology in particular? Some argue that the spatial aspect of information processing and analysis so well publicized in the GIS literature suggest that GIS is different from IS in general. Maybe these early experiences with information systems do not adequately predict similar results when GIS technology is applied to organizational computing. A look at some of the current research into successful GIS implementations may help to draw parallels and to recognize differences that will assist in understanding the important issues in GIS implementation.

Case Study 2. Correlates of GIS Adoption Success
(Pinto and Onsrud Study)

In 1991, questionnaires were sent to 506 local government users of GIS technology in the United States, Canada, Europe, Australia, Asia, and Latin America by researchers at the University of Maine (Pinto and Onsrud, 1991). The 256 usable responses were then analyzed to gain an understanding of the factors that were reported as most influential in the successful adoption and use of the technology by those users. The results indicate that six factors are most predictive of success:

- Advantages of GIS over current processes
- Ease of getting results
- Data accuracy adequate for the agency's needs
- Consistency with the goals of the agency
- The ability to expand the types of uses in the future
- History of past failures

The researchers conclude that users of successful GIS perceive that their GIS is valuable to their organization because it is easy to assimilate and use, and that their organization had experience with past computer system failures. This last factor, history of computer system failures, suggests to the researchers that an organization's concern with past computer systems failures helps ensure success of the GIS project. Perhaps those who have been "burned" in the past require more scrutiny in their GIS project than those who are not aware of the adverse impacts of computer system project failures. Further, the researchers conclude, a GIS is most likely to be viewed as successful when the users perceive that: the system provides more advantages than their old methods; the organizations have fallback options in the event that the system does not perform to their satisfaction; and the costs of the system are not excessive.

The research also determined that the successful users of GIS technology employed similar processes to acquire the GIS technology. The following 10 steps were identified as important by a majority of respondents (percentages indicate the percentage of respondents who reported that the step was taken):

- Enter a purchase contract (92%)
- Identify user needs (87%)
- Match GIS to tasks (85%)
- Seek staff support (85%)
- Prepare an informal proposal (82%)
- Prepare a formal proposal (79%)
- Prepare a request for proposal (79%)
- Identify GIS location in the organization (76%)
- Conduct a pilot project (69%)
- Acquire a GIS consultant (53%)

While it appears that there is still some question of the importance of using a consultant to assist in the GIS project (slightly more than half reported using one), it is clear that identifying user needs (87%) and matching the GIS design to specific tasks (85%) are considered critical to the eventual success of the project. In addition, it should be noted that a significant number of projects (69%) used a pilot project in their process of GIS implementation.

Case Study 3. Organizational Issues in Successful GIS Implementation (Campbell Study)

A successful GIS project proceeds through three distinct phases: adoption of the vision that GIS can improve the organization; implementation of the system in accordance with user needs; and, finally, use of the system once it has been implemented. While it is the successful use of the system that is the true measure of the success of a project, the project cannot reach that phase unless the concept is accepted and funded by decision-makers and until the necessary hardware, software, data, people, and procedures have been implemented.

These statements are supported by the research of Heather Campbell (1992) at the Department of Town and Regional Planning at the University of Sheffield. Campbell investigated nine agencies in Massachusetts and Vermont in 1990 that had adopted GIS technology and found that the most important phase of their projects was the process of implementation:

Effective utilization was not found to be simply dependent upon the technical operation of a GIS. Organizational issues, including the ownership and control of information, securing general commitment and ensuring the needs of users are met through a realistic understanding of the role [of] information in decision making, were found to have a marked influence on the implementation process. (p. 86)

Campbell's study involved interviews with GIS system designers and users in a range of local and state agencies who had successfully implemented GIS technology. The distinction between adoption and implementation of the technology is described as follows:

The decision to introduce new technology into an organization may be protracted but there is a finite point at which a decision is taken to commit at least initial funding. In contrast, implementation is an on-going process involving repeated cycles of development, learning, and routine utilization as new elements are integrated into the existing system. Each of these cycles entails the complex process of managing change in environments which are themselves dynamic and characterized by the interplay of individual personalities in a context of distinctive procedures and long standing practices. (p. 86)

The findings of Campbell's research indicate that issues change during the implementation process, initially centering around technical problems such as system compatibility, and then progressing to data-related issues such as lack of consistency between data sets. As progress continues, the issues became more organizational in nature, revolving around difficulties concerning the ownership and control of the geographic information stored in the system and how to ensure that the information will be used in the policy-making process.

The interviews suggest (in retrospect) that the effective utilization of these systems would have been aided by addressing the following issues early in the implementation process:

1. Organizations and units within them often jealously guard their scope of activity and treat with suspicion any proposal that implies a change to these circumstances.
2. Professional staff and decision-makers made limited use of GIS capabilities, especially in the case of decision-making related to environmental and planning related issues as opposed to the administrative uses associated with cost saving issues.
3. Local communities are highly suspicious of a development that suggests centralization of information and, therefore, power.
4. GIS specialists are uncomfortable handling the more social and political aspects of systems utilization, which implies that involving users early in the process of implementation can improve the effective utilization of GIS technology.

A Need for Management Perspectives on GIS Implementation

All of this research indicates that GIS implementation raises a series of management and organizational issues. The Onsrud and Campbell research from case studies of successful GIS implementations in the United States produced findings that strongly support the Kraemer et al. (1989) research on information systems in general. Successful GIS implementation projects appear to share some common characteristics: a shared understanding or vision of how the GIS will be used; a comprehensive approach to defining and managing data; a formal approach to the selection of enabling technology; and a management framework that facilitates systematic analysis and planning of GIS implementation.

Appropriately planned and managed, the data, technology, and human expertise required for GIS implementation and operation can be shared as organization-wide resources. Data can be collected, stored, and maintained once, to the benefit of all authorized users. Technology can be implemented so that it is modular, can be implemented in increments, and permits efficient communication between modules. Technical expertise can be effectively shared, reducing costs and supplying adequate technical support to everyone. The processes and procedures can be organized to provide all participants with management autonomy and a role in the shared system. However, the advantages will not be realized unless the implementation process is explicitly managed to produce them. There is a need for a guiding philosophy supported by policy and management guidelines, and an organizational structure to implement and monitor the policy and guidelines—a Management Framework (establishing the Framework is discussed in Chapter 3).

A Management Framework defines how the process to implement the GIS will be structured, funded, and monitored. Fundamental to the framework is a set of policies or guidelines for GIS implementation composed of principles that provide a basis for setting specific objectives and monitoring progress. Table 2.1 shows an example of a set of GIS principles drawn from examination of successful GIS implementations in North America by a joint project carried out by the International Association of Assessing Officials (IAAO) and the Urban and Regional Information Systems Association (URISA).[1]

These principles are independent of any particular technical or organizational structure and provide the basis for management to evaluate technical specifications and project results. The principles can be measured and are equally meaningful to technical and management staff.

If a policy framework is not established, it will be more difficult for management to set direction and evaluate progress. The problem with attempting to manage by monitoring technical aspects alone is that evaluation of progress becomes curtailed by the need to evaluate endless amounts of often difficult to understand technical detail. A set of guiding principles permits each technical aspect to be compared against a corresponding guiding principle. If the technical aspect complies with the guideline, then no concerns exist. If it does not, then management can explore the matter further.

Table 2.1   IAAO GIS Principles

Principle 1: A GIS is a data-driven, data-based information system. Making maps is only one capability of GIS technology. GIS is based upon database management concepts that allow flexible access to data, integration of data from different sources, and manipulation and analysis of data in ways that are not possible with other techniques or technologies.

Principle 2: GIS data and maps must be maintained. Because the GIS is designed for use in the operating environment of the organization for specific service delivery, management, and policy-related activities, the data and, ultimately, the system will not be used if they are out of date or if they are inaccurate.

Principle 3: A GIS is most useful when geographic references are registered on a consistent, continuous coordinate system. A GIS is not merely a collection of computerized map sheets. Its use demands that an entire geographic area be accessible in order that the spatial relationships among features in different parts of the area can be identified and displayed. This requires locating each map sheet in its proper place within the entire geographic area. It is accomplished by setting coordinates for each map sheet that come from a consistent, continuous coordinate system.

Principle 4: A GIS has topology. Since a computer cannot see a map as a human can, additional definitions of the relationships among the points, lines, and areas must be established. This topology allows the GIS to perform certain spatial analysis functions including (but not limited to): network analysis and optimal path determination, polygon overlay, geocoding, and area calculations and shading.

Principle 5: A GIS has many uses and should be shared by many different functions. Since the value of information increases the more it is shared and used by others who need it, and since a GIS requires a significant amount of resources to develop, a multiuse, shared system can prevent duplication of common data and the effort to maintain them and can also reduce the cost of the system to any single user.

Principle 6: A GIS contains hardware and software that are constantly undergoing change, which improves its functionality over time. A delay in acquiring GIS hardware and software in anticipation of future price reductions or technological breakthroughs is not prudent because existing technology is fully adequate to develop GIS applications. The benefits of the system cannot be realized until the databases are built and implemented on the system, so a delay in implementation only creates a delay in realizing its benefits. Future improvements to the technology will enhance the use of the data - not restrict it.

Principle 7: A GIS grows incrementally in terms of technology, cost, and administrative support needed. Therefore, a long-term commitment is needed to assure success. The large amount of time required to build the databases and the large number of potential users and applications prevent a GIS from becoming fully functional within a short time frame. Given limited resources and an ambitious plan for GIS implementation, priorities must be established and commitments maintained over a multiyear time frame.

Principle 8: A GIS causes changes in procedures, operations, and institutional arrangements among all users. The common databases accessed by many different users eliminates the compartmentalization of data storage and individualization of data coding schemes. This will result in changes in responsibilities, procedures, security measures, standards, and even organizational structures and laws in order for the GIS to function for the benefit of all users.

Principle 9: A cadre of trained, educated, motivated, and dedicated people is crucial to a successful GIS program. Without exception, organizations that have successful systems have been able to assemble, and retain for a long time, the appropriate level of staff with technical and communication skills who have, as well, a shared understanding and vision of the potential for the technology. Technical problems can be resolved with money and time, but staff without the motivation, dedication, creativity, and willingness to accept new ideas are likely to scuttle the project. Most successful systems have had a "champion" - a high-level official who is willing to push the project forward and motivate and educate those whose support is needed. However, increasingly, successful organization-wide GIS implementation is being led by a dedicated team, rather than a single "champion."

Management of the policies and principles requires an organizational structure that provides for active participation in determining what will be done, who will do it, and how the impacts of change will be managed. The intent of the organizational structure is to stimulate participation and thus enthusiastic support for the long-range implementation, not to create controlling mechanisms and compliance by participants.

The IAAO principles have been formed by studying the common character-istics of successful systems by a multidisciplinary team of experts—experts in information systems, surveying, photogrammetry, land assessment, economics, public administration, and geography and cartography. They have not been developed in isolation by a "GIS science," but rather from an empirical study of "GIS practice," which draws upon proven concepts from a variety of different disciplines.

In Chapter 1, an explanation of the GIS paradigm revealed that the founding concepts of GIS originate with many different land-related sciences. Similarly, many of the concepts, and subsequently the techniques and practices used to plan, design, implement, and manage GIS technology, originate in a body of knowledge that has its origins in information resources management and management science and practice. A brief overview of these topics is appropri-ate since much of the approach to GIS implementation that will be presented in this book builds on this work.

Information Resources Management

Information resources management (IRM) is a management philosophy that is founded on the belief that information is an asset that should be managed rigorously and that should contribute to the success of a organization. The concept treats all the information and computer systems in an organization as a package to be planned, managed, and controlled in order to be used most effectively and efficiently. Its basic premise is that the computing resources of an organization—the hardware, the software, the data, and the people—are interacting with each other on a daily basis and are constantly undergoing change, individually as well as collectively. IRM seeks to manage this change in an orderly, cost-effective, and appropriate manner.

IRM, then, aims to institutionalize the use of computing technology on an organization-wide basis in order to satisfy the information needs of the operations, management, and policy-level responsibilities of an organization. It also provides the assurance that the appropriate data and technology infrastruc-ture is in place when and where it is needed to ensure that the applications can, in fact, be used. This is an important distinction: The institutionalization of computing requires the appropriate design techniques for the use of data as well as the appropriate technology infrastructure for the delivery of the information when and where it is needed in an organization. Concepts of management information systems (MIS) and information technology (IT) planning are now being used by information resource managers to plan, manage, and control the package of information resources in an organization to assist in this institutionalization.

Management Information Systems

When computer technology was first introduced into organizations in the 1950s and 1960s, it was successfully used in applications designed to automate the procedures of processes that had existed for many years: processes such as payroll, accounting, personnel, etc. The resulting "process-oriented" design had as its central focus a "transaction" (e.g., a timecard or an accounting entry) that triggered specific tasks that had once been performed manually. During this era, computer technology was referred to as "data processing" because computer programs processed data from transactions for use in processes.

Geographic information systems had their origins in this data processing era, although the systems used at the time were not known as geographic information systems. Computer graphics systems were developed during this time to support many graphical processing activities of organizations: drafting, mapping, and designing. These systems soon became known as computer-aided drafting (CAD), computer-aided drafting and design (CADD), computer-aided mapping (CAM), and automated mapping (AM) because they automated the manual processes of drafters, designers, cartographers, and engineers. They could be used to create, modify, store, and retrieve the maps, drawings, and plans used by these professionals.

In the 1970s and 1980s, commercial database management systems (DBMS) were developed and implemented in these organizations in order to help manage the large databases created by these data processing systems. These DBMSs also allowed system developers to design systems for more than one application—a series of different computer programs that used the same "database" for more than one application. Thus the focus of design changed from transaction-oriented processing, supporting existing procedures and responsibilities within the organization, to information-oriented processing, supporting a wider set of applications useful for many different processes and levels of responsibility in the organization. These new systems, using database management systems software, soon became known as "information systems" that not only were used by the clerical staff in the daily processing of specific tasks, but also became valuable for higher-level requirements of managers in support of a variety of resource management responsibilities. This new design, a comprehensive database,[2] or integrated set of databases in support of many different applications used by more than one function in the organization, became the rage of the information systems field and soon became known as "management information systems" because they supported these higher-level information needs of managers.

These database management systems also allowed the computer graphics systems of the earlier era to be expanded in their applications, just as data processing systems expanded into management information systems. Databases containing information related to geographic locations soon became part of computer graphics software. The resulting technology, *geographic information systems*, provided the capability to combine location-related data with digital images of maps for a wide range of applications to support the geographic information needs of planners, managers, administrators, and policy-makers. No longer was the process-oriented computer graphics technology the exclusive domain of the drafters and designers—GIS technology became an organization-wide information resource just as MIS technology did when DBMS hit the data processing scene.

This discussion on the history of the use of computers and its similarity to GIS technology is presented in order to highlight the parallel between MIS and GIS when the technology is moved out of an individual function (a process) into a more comprehensive, multidisciplinary set of applications. As will be seen in the later chapters of this book, existing institutional frameworks that organize work around rigid structures based on work and information flow of earlier computing eras are most likely to prevent the successful implementation and use of the new technology. It is therefore important for investigators of GIS technology to follow trends closely in the larger MIS field (given its 30-odd years of development history) when seeking successful implementation and use of GIS.

Information Technology Planning

Many organizations are still using the process-oriented systems designed years ago. Since the commitment to organization-wide management information systems requires a significant amount of resources, time, and involvement of many people, the tendency is to think: "If it ain't broke, don't fix it!" So change is slow. New applications are generally minor in nature, and usually consist of modifications to existing systems. Thus it is difficult to move from a computing environment with individual systems developed in isolation and with redundant databases that cannot be shared among different units of the organization to management information systems that integrate databases for use by many functions and by different levels of the organization.

Those organizations that have moved on to an organization-wide computing environment have not found the change easy. The motivation to change has usually been precipitated by top management awareness of the rising costs of computing, coupled with user department complaints that their computer systems do not perform in accordance with their changing needs. As the data processing department spends a majority of its resources on modifying existing systems, few, if any, new applications become implemented.[3] These organizations found that this emphasis on current system maintenance was stifling the growth of data processing and that there was no way to look to the future for improvements. Martin and McClure (1983) recommend that management should get involved when this situation occurs:

Unless radical management action is taken, such companies will remain stuck indefinitely, and the business ramifications may be disastrous. The only way out is migration to new technologies. Management must be resigned to cut its losses and to make a commitment to move on. A careful plan must be devised to determine which new technologies will be most useful, to train DP personnel and end users in the new technologies chosen, to reorganize DP, and to control future growth. The plan should provide long-term goals against which DP progress can be measured. It should also specify the immediate actions necessary to get things moving. Finally, management must implement the plan and remain committed to it. (pp. 447–50)

Further, Martin and McClure insist that the plan must be compatible with the overall business plan of the organization, and that it must have realistic expenditures and application areas that are most critical for achieving corporate goals.

This high-level approach to information technology planning (stressing compatibility with the overall business plan of the organization) allows for the strategic planning of how IT should evolve in an organization. It provides a general structure that allows individual modules to be developed over time by separate teams so that pieces of the different information systems can fit together where they must. These modules will not fit together unless they are designed with planning from the top and with good design tools.

Management Science and Practice

It is now widely recognized that successful information technology implementation and use depends as much on successfully dealing with management and organizational development issues as it does on high-quality systems design. Many management issues will be raised in the course of the discussions in this book. The most relied-upon areas of study providing guidance with these issues are organizational theory (or organizational development) and management practice as it relates to the management of people, money, and technology.

Organizational Theory

Increasingly it is becoming recognized that the implementation of information technology cannot be separated from other aspects of the organization into which it is to be implemented. This is at least as true for GIS as it is for any other information technology application, perhaps even more so. Successful GIS implementation therefore includes successfully dealing with characteristics of organizational culture, the dynamics of people interacting in teams, change processes, and the impacts of introducing new technology. It is this last aspect in particular that will be focused upon, relying upon the body of knowledge that has grown around the management of technology in organizations and technology-related change. Of particular significance are the works of Shoshana Zuboff, Paul Goodman and Lee Sproull, Peter Keen, and Richard Walton.

The specific issues related to technology in organizations have been combined with broader perspectives of the changes happening to organizations generally in works by management leaders such as Peter Drucker, Tom Peters, and Rosabeth Moss Kanter. These supplement the more fundamental works in organizational behavior as represented by Paul Hersey and Kenneth Blanchard, Robert Guest, Richard Beckhard, and Edgar Schein.

Throughout these works some common themes emerge:

1. There are fundamental changes occurring to the management of organizations around the world, and the increased use of information technology is both a cause and an effect of those changes.
2. The people-oriented aspects of organizational change are both more important and more challenging than the technological aspects.
3. The application of information technology can contribute significantly to improving the performance of organizations, but an information technology application, such as GIS, alone will not transform an organization.
4. Successful implementation (systems that are used and that improve the performance of organizations) and high returns on investment in information technology depend on a planned, well-conceived and -managed integration of information technology and organizational change.

The reader will find these themes applied and re-emphasized in the following chapters.

Management Practice

Many organizations are now facing the challenges of building an electronic enterprise. The process involves reinterpreting the fundamental business activities of the organization, applying information technology to them, and managing the associated changes. Prerequisite for success is the effective use of people, including: securing their commitment to support the changes, building and growing effective teams to perform the work, and creating effective working environments for such teams. The human aspects must be managed in conjunction with well-conceived planning and management frameworks.

As information technology increases in sophistication and the per unit cost decreases, more of an organization's business activities become computerized. However, people remain the key to successful endeavors. The two following lists show the aspects of work that may be computerized and those that will not (Martin, 1984, p. 265).

Aspects of work that will be automated:
  • Simple tasks
  • Routine tasks

- Reproducible tasks
- Tasks requiring complex logic or calculation
- Expertise that can be reduced to rules
- Tasks requiring vast knowledge

Aspects of work that will not be automated:
- Originality
- Intuition
- Inspiration
- Art
- Leadership
- Salesmanship
- Humor
- Love
- Friendship

Not only will the aspects in the second list not be automated, but many of the qualities are also essential to successful business endeavors. Nurturing these qualities in individuals and groups requires appreciation of the complexities of human individuality, the dynamics of group behavior, the impact of working environments, and the affect of the attitude of management and co-workers.[4]

GIS implementation (as with other IT implementation) will facilitate higher productivity from knowledge workers. Peter Drucker argues that the single greatest challenge facing managers in developed countries of the world is to raise the productivity of knowledge and service workers. This challenge, which will dominate the management agenda for the next several decades, will ultimately determine the competitive performance of companies, the very fabric of society, and the quality of life in every industrialized nation (Drucker, 1991, p. 69). This significant observation has two impacts on our discussion: The first is the need to apply the capabilities of GIS appropriately to land-related work to improve worker productivity; the second is effectively to improve the performance of GIS implementation teams to reduce development time, decrease costs, and accelerate benefits from GIS implementation. Applying the best of new and proven management techniques to GIS implementation provides part of the answer to required productivity gains.

Another closely related factor is the issue of management participation in the application of information technology and the associated changes. Managers are developing a better understanding of the significance of information, and the technology to utilize it, as they realize the value of their information assets and the cost of failing to exploit them fully. Keen (1991) identifies three issues in managing the economics of information capital: managing costs, managing benefits, and managing risk exposure. He then observes that these have generally been handled poorly because IT has been treated as an overhead, managed through budgets, cost allocations, and cost justification. This is a very naive way of dealing with a complex economic good.

To succeed with GIS implementation (i.e., fully exploiting its potential and the organization's investment), a strategy and management framework that establishes GIS and geographic data as a strategic asset is needed. Included are: involvement of senior management in establishing policy and direction, aligning business functions and technology strategy, viewing technology investment as a strategic investment, and fostering commitment and competency in the organization.

Improved strategic management of information resources needs to be coupled with effective project management for strategic initiatives to realize promised benefits. The principles and techniques for management of complex engineering projects have been known for some time. Many of these date back to major military and aerospace projects such as construction of the POLARIS submarine, for which Project Evaluation and Review Technique (PERT) was developed.[5] These techniques are now available in a variety of software packages costing from a few hundred to several thousand dollars. Additionally, training in both the use of the packages and project management principles is readily available from private training institutes and a host of college extension programs.

Well-planned design and construction must be balanced with the realities of the backlog of applications, ever more competitive business environments, and declining budgets. This reinforces the need for efficient streamlined project management that can balance emphasis on quality assurance with early returns on investment.

Grayce Booth (1983) suggests six ground rules for successful design of complex computer-based systems:[6]

1. Involve the people who will use the system, as well as the management to whom those people report, in the process of system analysis and design.
2. Define the goals and objectives of the system as completely as possible before beginning design.
3. Obtain agreement, in writing, from the users and their management that the defined goals and objectives are appropriate.
4. Do not view the system objectives, even when agreed to in writing by users and their management, as cast in concrete.
5. Realize that prospective system users, unless they are very familiar with computer-based systems from past experience, are often unable to visualize what such a system can do. In that case, measures to raise their awareness should be considered to facilitate meaningful participation.
6. Organize the implementation and installation of the system in manageable phases with formally defined checkpoints, at each of which specific tasks must be accomplished. Too many large projects are undertaken with the idea that the system must be implemented as a whole, even if this takes several years. Booth found in one instance that implementation

was begun with an expectation that there would not be usable results for 8 years, during which no organization's situation can remain stable enough for such a time frame to be rational.

Booth observes that these ground rules have very little to do with technology or with methodologies for computer system design. They are concerned with communication and relationships between technical staff (system designers) and users and their management. Booth's experience indicates that system failures are most often caused by failing to define exactly what the system ought to do, and she stresses proper communication as a central theme in successful IT implementation.

Conclusion

The intent of this chapter has been to broaden the context for GIS implementation. The three case studies and the review of contributions from information resources management and management practice have been included to establish a foundation for the discussion of GIS planning, design, and implementation in the remaining chapters of this book.

In summary, the foundation that will be built upon in the remaining chapters consists of the following premises:

1. GIS comprises a set of concepts for organizing work and data (as defined by the GIS paradigm) as well as enabling technology.
2. GIS is a complex information technology application and as such adheres to many of the same tenets and is affected by many of the same issues as other complex information technology applications.
3. A comprehensive, systematic approach to planning, design, and implementation management will produce a more successful GIS implementation.
4. Management and institutional issues compose a large portion of the factors that determine successful implementation and consequently require careful consideration during implementation.
5. Organizations that create open, participative GIS projects are more likely to deal successfully with the many management, institutional, and technical issues that comprise GIS implementation.

NOTES

1. These nine "GIS principles" emerged during meetings of the Joint IAAO/URISA Committee on GIS Guidelines for Assessors, consisting of experts from the International Association of Assessing Officers (IAAO) and the Urban and Regional Information Systems Association (URISA). After several initial meetings, during which the "best" methods for designing and implementing a geographic information system were debated, the experts were deadlocked, unable to agree on whether such guidelines should

recommend the best methods, given the diverse needs and environments of the many assessing jurisdictions they represented. Specific recommendations, such as using digital orthophotography for construction of the base map as opposed to other methods, using a relational database management system as opposed to an hierarchical database management system, and using the North American Datum (NAD) of 1983 as opposed to the NAD 1929 Datum, could not be unanimously agreed upon by the joint committee because some jurisdictions did not even use maps, others had no computer or surveying expertise, and many did not have the financial or other resources required to design and implement the "best" system. The deadlock on recommendations was broken when the committee stepped back from the specifics and individuals began to discuss the bases for their points of view. It soon became apparent that they all agreed on certain characteristics of an "ideal" system, but when it came to the technical methods used to achieve it, consensus broke down. These agreed-upon characteristics of an ideal system became the GIS principles because the committee acknowledged that the conditions under which the technology is implemented differ from agency to agency. Further, future improvements in technology, information systems methodologies, and management practice could render any specific standard obsolete. Thus, for example, while there is debate on whether the best system contains a relational database versus a hierarchical database, all could agree that "A GIS is a data-driven, data-based information system (Principle 1). The specific type of database management system used depends upon the requirements and capabilities of the agency implementing the system. The remaining principles reflect similar generalizations about the "ideal" system.

2. Martin (1976) describes a database as a collection of interrelated data stored together with controlled redundancy to serve one or more applications in an optimal fashion; the data are stored so that they are independent of programs that use the data; a common and controlled approach is used in adding new data and modifying and retrieving existing data within the database.

3. The data processing management in the City of Milwaukee found in 1990 that 80% of the time reported by its programmers and analysts was spent on maintaining existing systems, and that there was a 3-year backlog in requests for new systems by user departments.

4. For an excellent text on improving moral and the productivity of people, see DeMarco and Lister (1987).

5. More information on PERT and related techniques can be obtained in Weist and Levy (1977).

6. Booth (1983) defines a complex information system in terms of their infiltration into and impact on the organization, integrating computer-aided system (information technology) with business functions, across the organization. GIS can also be considered as a complex application of information technology for many of the same reasons. The various data types associated with geographic data (line cartography, land statistics, textual records, analog images, digital pixels) must be integrated, usually from several disparate sources, frequently from outside the organization. Spatially referenced data call for specialized tools for data capture, maintenance, and presentation (i.e., maps). The technology employed is varied, not fully integrated, and complex. Much of it is still being developed, and new technology is being added constantly. Very large databases are often involved. Geographic databases are some of the largest databases ever to be developed. Additionally, for a variety of reasons, they are often distributed across the organization, requiring complex networks and software to link and maintain them.

REFERENCES

Booth, G. M. N. (1983). *The Design of Complex Information Systems: Common Sense Methods for Success.* McGraw-Hill, New York.

Campbell, H. (1992). Organizational issues and the implementation of GIS in Massachusetts and Vermont: Some lessons for the United Kingdom. In *Environment and Planning B: Planning and Design,* vol. 19, pp. 85–95. Pion, London.

Drucker, Peter F. (1991). The new productivity challenge. *Harvard Business Review,* November–December.

GIAC (1992). Towards a shared vision of the future of shared vision of the future of geomatics in Canada. In *A Consultation Session to Develop a Competitiveness Plan for the Canadian Geomatics Industry,* Geomatics Industry Stakeholders Seminar.

Goodman, Paul S., and Lee S. Sproull (1990). *Technology and Organizations.* Jossey-Bass, San Francisco.

Huxhold, W. (1993). The application of research and development from the information systems field to GIS implementation in local government: some theories on successful adoption and use of GIS technology. In *Diffusion and Use of Geographic Information Technologies,* Ian Masser and Harlan J. Onsrud (eds.), chap 1.4. Kluwer, Deventer.

Keen, Peter (1991). *Shaping the Future: Business Design Through Information Technology.* Harvard Business School Press, Boston, Mass.

Kraemer, K. and J. King (1979). A requiem for USAC. The Regents of the University of California, Irvine, California.

Kraemer, K., J. King, D. Dunkle, and J. Lane (1989). *Managing Information Systems: Change and Control in Organizational Computing.* Jossey-Bass, San Francisco.

Martin, J. (1976). *Principles of Data-base Management.* Prentice-Hall, Englewood Cliffs, N.J.

Martin, J. and C. McClure (1983). *Software Maintenance.* Prentice-Hall, Englewood Cliffs, N.J.

Martin, James (1984). *An Information Systems Manifesto.* Prentice-Hall, Englewood Cliffs, N.J.

Nolan, R. (1979). Managing the crisis in data processing. *Harvard Business Review,* March–April.

Pinto, Jeffrey K., and Harlan J. Onsrud (1991). Correlating adoption factors and adoption characteristics with successful use of geographic information systems. In *Diffusion and Use of Geographic Information Technologies,* Ian Masser and Harlan J. Onsrud (eds.), chap. 2.4. Kluwer, Deventer.

Urban and Regional Information Systems Association and the International Association of Assessing Officers (1992). *GIS Guidelines for Assessors.* International Association of Assessing Officers, Chicago.

Weist, Jerome D., and Ferdinand K. Levy (1977). *A Management Guide to*

PERT/CPM: With GERT/PDM/DCPM and Other Networks. Prentice-Hall, Englewood Cliffs, N.J.

ADDITIONAL READINGS

Aronoff, S. (1989). *Geographic Information Systems: A Management Perspective.* WDL Publications, Ottawa.

Baumgartner, John S. (1986). *Project Management.* Irwin, Homewood, Ill.

Beckhard, Richard (1969). *Organization Development: Strategies and Models.* Addison-Wesley, Reading, Mass.

DeMarco, Tom, and Timothy Lister (1987). *Peopleware: Productive Projects and Teams.* Dorset House, New York.

Finch, F., H. Jones, and J. Litterer (1976). *Managing for Organizational Effectiveness: An Experimental Approach.* McGraw-Hill, NewYork.

Gotlieb, C. C. (1985). *The Economics of Computers: Costs, Benefits, Policies, and Strategies.* Prentice-Hall, Englewood Cliffs, N.J.

Guest, Robert H., Paul Hersey, and Kenneth H. Blanchard (1982). *Organizational Change Through Effective Leadership.* Prentice-Hall, Englewood Cliffs, N.J.

Hersey, Paul, and Kenneth H. Blanchard (1982). *Management of Organizational Behavior.* Prentice-Hall, Englewood Cliffs, N.J.

Kanter, Rosabeth Moss (1983). *The Change Masters.* Simon and Schuster, New York.

Kanter, Rosabeth Moss (1989). *When Giants Learn to Dance.* Simon and Schuster, New York.

Metzger, Phillip W. (1981). *Managing a Programming Project,* 2nd ed. Prentice-Hall, Englewood Cliffs, N.J.

Peters, Tom (1987). *Thriving on Chaos: Handbook for a Management Revolution.* Knopf, New York.

Rosenberg, N. (1982). *Inside the Black Box: Technology and Economics.* Cambridge University Press, Cambridge.

Sheen, Edgar H. (1969). *Process Consultation: Its Role in Organizational Development.* Addison-Wesley, Reading, Mass.

Walton, Richard E. (1989). *Up and Running: Integrating Information Technology and the Organization.* Harvard Business School Press, Boston, Mass.

Zuboff, Shoshana (1988). *In the Age of the Smart Machine: The Future of Work and Power.* Basic Books, New York.

3

Strategic Planning for GIS

The beginning is the most important part of the work.

PLATO

Organizations around the world have embarked on the development of GIS systems. These organizations include local government, state and provincial government, agencies of national government, forest products companies, oil companies, marketing firms, railways and other transportation companies, health care agencies, and many others. Marketing studies estimate that by 1994 organizations will spend $3.4 billion per year on GIS hardware and software (Hughes, 1990). George E. Brown, Jr., Chairman of the House Science, Space, and Technology Committee, speaking at GISDEX in Washington in 1991, noted that 95 agencies in the U.S. federal government currently use GIS.

As new uses of the technology are discovered, analysts predict that the GIS market will grow to worldwide revenues of $21 billion by fiscal year 1997. Recent developments have seen GIS systems applied to everything from the legal arena to the marketing environment. In the private sector, GIS systems, now available on networks, are becoming integration platforms for various sets of corporate database information, with implications for the retail as well as the energy industries (Computing Canada, 1992). In the corporate market, GIS programs are used to incorporate maps with demographics, taxes, and various other types of numerical information that is usually figured in tabular form. Managers are about to combine the information from the maps and numbers in different ways in order to make "what if" comparisons. Various relationships may also be uncovered that were not apparent when the information was in tabular form. Its popularity has grown due to lower cost and the increased performance capabilities of microcomputers and workstations. The cost of GIS software has also been reduced and runs from $500 to $30,000.

Why are so many organizations investing in GIS? If we exclude the lemming theory ("because the other guy is"), then the simple answer is that the early success stories of the 1980s have demonstrated that implementing a GIS will improve the efficiency and effectiveness of their organizations. They want to save money, reduce staff, or perform an increased level of service without increasing staff, and they want to perform new capabilities that were not practical without a GIS. In short, they hope to achieve benefits in excess of what it will cost them to implement the GIS.

Further fueling the introduction of GIS technology is the general revolution in information technology. Computers are becoming cheaper, smaller, and easier to use, and more organizations are simplifying their purchase. With cheaper technology and the rapid emergence of new GIS tools, GIS benefits should abound.

The reality is, unfortunately, somewhat different. A 1989 study of GIS installations in Canada and the United States found relatively few GIS installations that can be considered fully successful (Croswell, 1989). Moreover, this situation is not unique to GIS, but rather is a characteristic of most complex information systems.

Most complex information system projects do not fail because of a lack of technology or because of technical failures. A comprehensive study involving case histories for about 500 system development projects since 1977 found that 15 percent of the studied projects were aborted in progress or delivered products that were never used (DeMarco and Lister, 1987). Twenty-five percent of major projects (involving 25 work-years or more) failed to be completed (Jones, 1981). In the overwhelming number of these projects, there were no technological issues to explain the failures. Institutional (sociological) reasons were almost always the cause. The Campbell (1992) study in Vermont and Massachusetts confirms the significance of the implementation process. Effective utilization relies on attention to organizational issues, including: the ownership and control of information, securing general commitment, and ensuring that the needs of users are met through a realistic understanding of the role of information in decision making (Campbell, 1992).

Why have so few organizations realized the benefits that should have been forthcoming, and how can greater success be achieved?

In this chapter it will be argued that a comprehensive, planning-based approach to GIS design and implementation is required to derive the order of benefits that have been promised by vendors, consultants, and at conferences.

Many GIS implementations have started out as back-room functions in either the drafting room or as the project of an individual technology advocate. Management information system (MIS) departments often left the GIS folks to themselves, because they did not have expertise with GIS and because GIS was not seen as being mainstream to the organization. Also, the MIS group was usually wrestling with application backlogs and the rapid changes to mainstream information technology (IT) applications themselves.

Early GIS applications focused on automating map drafting and basic geographic queries, usually within single departments. In cities, engineering and planning departments have been the main proponents. Many smaller municipalities did not have an MIS or equivalent group. Engineering departments in cities captured utilities data, but frequently did so without much consultation with the planning or assessment departments, who (if they were using any GIS technology themselves) were capturing their own data to different levels of accuracy and on their own standalone machines (often of a different brand and design from the GIS in engineering). Forestry companies and government natural resource organizations digitized resource inventories and performed map coloring and various locational analyses. Data were generally loaded on a project specific basis, without extensive database design or consideration for how the data would be maintained on an ongoing basis and/or for additional purposes.

Despite the somewhat fragmented approaches to GIS in most organizations, adequately successful examples of GIS implementation exist to demonstrate some benefits of GIS and extend the promise of others. At its most basic level, computerization of the mapping process yields glitzy and immediate results.

Barring complete incompetence, implementation of GIS as a basic support tool will produce benefits. Various studies have indicated that 1:1 to 2:1 returns on investment can be achieved. That is not disputed. What is disputed is that the scope of benefits could have been larger, that those or even greater benefits can be shared by parts of the organization in addition to the drafting room, and that such benefits can be sustained as the organization goes through additional changes of structure and technology.

Organizations that implemented GIS on a departmental or single project basis are now finding it difficult to benefit from higher-level applications that require more data integration and a more detailed understanding of the domain areas to be supported. As GIS applications become more complex, more of the required data originate elsewhere, inside or outside the organization. On a single project basis, these data are easily enough converted and merged with the department's own data. As these requirements become more routine, the demands on the GIS group increase. However, without mechanisms for regular data maintenance and exchange, each GIS product continues to be a new and separate project. Eventually the GIS group will collapse under the weight of its own success.

As GIS moves from being single to multiple user and as the applications become more complex, integrating data and processes across the organization, a number of issues begin to emerge:

1. Previously isolated computer systems need to be connected in order to share data among different functional areas.
2. The ongoing, routine data maintenance and system support activities require a high level of resource commitment.
3. A better understanding of information flows in the organization is necessary to coordinate and integrate the requirements of the various departments that share data and business processes.
4. Higher levels of staff competency are required to develop and operate GIS applications.
5. The ability to benefit from advances in technology becomes difficult without throwing out existing technology and data.

The apparent solution to dealing with these issues is to develop a plan, perhaps a strategic plan. Although planning is not equally popular in all organizations, at some stage of complexity, agreement to develop a plan is reached. There is usually little argument that planning is important, but the question of how much is the right amount generates debate. When should it be performed and who should be involved? What process should be used? In order to answer these questions adequately a better understanding of the relationship between the organization and GIS is required. Some additional questions might be asked:

1. What kind of an organization and situation do we work in?
2. What kind of a GIS project would work best in the organization?
3. How quickly can our organization adapt to new technology?
4. What is an appropriate, manageable rate of implementation?
5. How can we best determine the strategy to use?
6. How should the planning exercise be conducted?
7. Who should be involved in and lead the planning exercise?
8. How can the approach be sold to others?

Answering these questions is the subject of this chapter. The balance of this chapter will discuss the relationships within the organization as it exists, factors that affect efficiency and effectiveness, the management of change, and the relationship of these considerations to GIS.

The premise that will be argued is that, through effective planning, the scope of benefits can broadened, and benefits can continue for a longer period with less organizational trauma. Success requires consideration of many aspects including organizational and institutional issues, the functional working of the organization, data use and management, and GIS technology and how it is introduced, planned for, and managed.

Successful design and implementation of GIS is very much related to dealing successfully with all of these aspects. And not just dealing with each aspect individually, but with the interdependencies between them as well. That requires having a well-conceived approach for proceeding.

The planning approach depends significantly on the characteristics of the organization into which the GIS is being implemented. We will begin by briefly looking at organizations, how they function, and why a GIS might be developed.

GIS and the Organization

Both public and private organizations exist to satisfy specific purposes, or *missions*. Private enterprises usually have missions related to the profitability of their owners and they develop *goals* to support that mission in the form of a business plan that defines products or services, markets, and financing opportunities. Public organizations, on the other hand, usually have missions related to the viability of living conditions of the people within their jurisdiction,

FIGURE 3.1 Organizational framework within which information processing occurs.

and they develop *legislation* and *mandates* to fulfill these "social contracts." Both enterprises, however, create *organizational structures* designed to fulfill their goals in pursuit of their mission and to place staff within those structures to implement *programs, policies,* and *procedures* for conducting the business of the organization.

While their missions generally remain constant over time, *driving forces* inside and outside the organizational environment change, causing their goals, structures, and operating mechanisms to be adjusted accordingly. Organizations, then, are dynamic environments whose goals, structures, programs, policies, and procedures are adjusted in response to changes in internal and external driving forces in order to achieve missions that remain fairly constant. How dynamic these organizations are depends upon whether they are in the public or private sector, their management philosophy, their organizational culture, and the extent of external change that exists in the world. This simplified description of the organizational environment of an organization, as shown in Figure 3.1, defines the framework within which information processing in the organization occurs.

Historically, organizational structures were created primarily to deliver products and services (design, production, marketing, etc. in the private sector, and transportation, housing, public works, health, public safety, etc. in the public sector). As data about the organization and its operating environment (customers, suppliers, citizens) become more important in day-to-day decisions, an additional responsibility of the operation of the organization has been added: to create a means for processing information—recording it, retrieving it, summarizing it, and simplifying it for use in decision-making. Thus the responsibilities of bureaucratic hierarchies, designed primarily for the chain of command responsibility in decision-making to produce a product or deliver a service, have evolved into a systemic mechanism for synthesizing information—from the details of business operations to the information needed to manage resources to the knowledge required to set policy and plan for the future of the organization.

The role for middle management in most organizations now is to synthesize and distribute information: upward distribution to top management for decision-making, and downward distribution to workers to direct operations. What organizations have been experiencing is that information has become another resource that must be managed. As a result, they have created information-handling mechanisms, filing systems, and reporting rules and procedures to assist in managing their information. Government agencies, resource companies, utility companies, and others created drafting and cartography departments to compile and display the geographic information needed to run the organization. Various resource and land use professionals have designed and managed processes for collecting, analyzing, and presenting geographic analyses. As these "information-oriented" functions have been added to existing "product- or service-oriented" organizational structures, a certain amount of pressure has been experienced on the way that work is accomplished in these organizations. While their missions remain the same, this organizational pressure has affected the organization's ability to deliver products or services effectively and efficiently.

Prior to the use of computers for processing information, people and physical files have performed these functions. Now, with information systems technology, and GIS in particular, new methods for processing valuable information have been developed. These new methods have ranged from the relatively straightforward automation of existing tasks to sophisticated data integration and analysis capabilities. The extent to which an organization adopts the new technology—whether it be merely to automate manual tasks or whether large-scale improvements in information processing capabilities are de-sired—determines the amount of pressure, or change, that will be experienced on the structure of the organizations.

The significance of this point is that, for many years, organizations have used geographic information handling processes and systems. All public organizations and most private organizations have existing organizational units that are staffed with people who are accustomed to performing information handling tasks in traditional ways. With the computer revolution, however, new computer-assisted tools now must be used by these people, and the extent to which they are integrated into organizational structures has a great deal to do with their success.

This helps us to understand why GIS technology in the form of automated mapping or computer-assisted drafting (CAD) systems is often introduced into the drafting department as the starting point for introducing GIS technology into an enterprise. Automated mapping can be implemented without significantly changing existing procedures, policies, or programs of the organization because it essentially automates an existing procedure: When part of an enterprise requires a map, they request it from the drafting department and drafting personnel produce it. Whether the drafting personnel use a computer or manual methods, the user (the requester) does not notice much of a difference. As recently as 10 to 15 years ago the technology did not support much more than this CAD operation.

With the creation of spatial databases and the production of both maps and reports from these databases by individual resource professionals without the aid of the drafting department, the impact of GIS technology to the organization has become much greater. Over time, unless Draconian measures are taken to prevent individuals from learning and using the technology, changes to the traditional workings of the organization will also occur. As staff begin to use the databases and GIS technology to answer their own questions directly, the traditional information flows in the enterprise begin to change. This change is certain. What is less certain, however, is whether these changes will be improvements.

The diffusion of GIS throughout the organization and the promise of greater benefits drives GIS technology into higher-level activities in the organization. As a result, improvements in efficiency and effectiveness, coupled with new capabilities to cope with greater complexities of the organization, become the motivation for buying GIS technology. However, these benefits can be elusive. If the GIS technology causes disruption of existing information flows without creating new equally well-integrated information flows, the benefits cannot be realized. The first example in the Appendix illustrates problems experienced by the government mapping agencies in Canada.

In introducing GIS technology into an organization, the accompanying enterprisewide paradigm shift[1] is far more significant than the computerization of individual information processing mechanisms. This paradigm shift means that, in order to realize fully the benefits of GIS technology, an enterprise must adjust its organizational structure so that it matches the capabilities of the technology.

There is a well-understood relationship between benefits and how smoothly and effectively the transition from analog to digital is made. In studying this issue we can borrow from research done by industrial process specialists studying the introduction of CAD/CAM technology in manufacturing organizations. They have identified clearly defined relationships between technology and structural arrangements. They have identified three significant variables that must be dealt with in matching structure to technology:

- *Complexity,* referring to the number of different items that must be dealt with at any given time by the organization
- *Uncertainty,* referring to the variability exhibited by materials and or work procedures or to the extent to which it is possible to predict what problems are to be encountered or what procedures are to be carried out
- *Interdependence,* referring to the extent to which the items or elements involved or the work processes are interrelated so that changes in the

state of one element affects the state of the others (Scott, 1987, pp. 212–14)

The greater the diversity, uncertainty, and interdependence associated with performing a given type of work, the more complex the process must be to successfully introduce the new technology.

Advanced forms of information technology such as organization-wide applications of GIS have broader and deeper organizational consequences than earlier forms. Increasing the functionality of an information technology increases the levels of learning and adjustment required to utilize it, ranging from operator skills to organizational procedures and structures, and to cultural fabrics. The more sophisticated applications change not only the mechanisms for information handling, but organizational structure and job functions as well.

Success with early basic applications of GIS technology often encourages GIS managers to proceed with more complex applications. Senior management supports investment in the new applications based on success with past applications. However, as the complexity and breadth of the applications increase, a new set of rules for success comes into being. The GIS technology begins to integrate separate organizational functions in new ways. These more complex GIS applications cut across existing lines of authority, and the GIS applications create change or the need for change. Such change is often not anticipated, or when it is foreseen, the required change may not be within the purview of the GIS manager and thus may be counter to the political environment of immediate superiors.

In these circumstances, implementing GIS technology is introducing a new way of doing business. It creates demands for changes to present business processes and results in a new way of looking at the organization's functions—the paradigm shift. In most cases the effects of the paradigm shift are more dramatic than the direct results of the GIS applications. However, GIS technology is the tool for implementing the paradigm shift, not the driving force. The paradigm shift eventually impacts all aspects of the organization. Successful implementation must therefore deal with all aspects of organizational impact. The organization must integrate change planning and technology planning.

Four tightly interconnected elements of GIS implementation were identified in Chapter 1: the GIS paradigm, GIS data management, GIS technology, and GIS organizational strategy. Each of these elements of GIS implementation is complex in and of itself. Additionally, to create a completely successful GIS implementation, they cannot be managed in isolation; they must be managed together, and various aspects must be aligned to achieve maximum benefits. Strategic planning creates a framework for managing change and the interdependencies between the technology, the data, and the organization and its people.

Strategic Planning: A Management Framework

A strategy provides a long-term view and hence the ability to take risks
or do things which do not make sense in the short-term.

EDWARD DE BONO[2]

The purpose of strategic planning is to create a framework within which the complexity and interdependency of GIS design and implementation can be managed. The process of creating the strategic plan merges the four elements of GIS with the workings of the organization in a common planning and management framework. Figure 3.2 illustrates this framework.

In the process of strategic planning, each of the four GIS elements is dealt with in the context of the existing organization. Individual GIS elements are dealt with in varying detail at different stages during GIS implementation, but all elements are collectively managed within the framework of the strategic planning process.

As discussed in the previous section, the scope of benefits from GIS implementation is governed by how well all aspects of implementation are managed and coordinated. These aspects are summarized in Figure 3.3.

The purpose of strategic planning is sometimes presented as being able to predict and control how the future will evolve. Anyone who has participated in strategic planning recognizes that view as naive. The real benefits are derived from integrated management of the various aspects of organizational structure, staff behavior, and information technology application, creating a solid basis for dealing with the unexpected. In the absence of strategic planning that forces proactive consideration of all aspects of implementation, success happens only

GIS Strategic Plan

FIGURE 3.2 The GIS strategic planning framework.

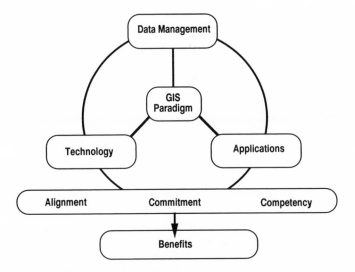

FIGURE 3.3 Scope of GIS benefits.

by chance. It is somewhat like driving a car through the rear-view mirror. Everything seems fine until the road turns suddenly, or the car collides with something. If the road is straight and clear of obstacles, a long time may elapse before any disaster occurs.

Richard Walton (1989) states that effective use of information technology depends on

- Good direction, embracing *alignment* of business, technology, and organizational strategies
- High organizational *commitment,* stakeholder support, and ownership
- Strong task *competence* in design and implementation management and user mastery of the technology

Information technology is a tool that, when applied effectively, can provide high levels of support to organizational processes. The degree to which information technology is supported drives the scope of benefits to the organization. Injection of GIS into the organization can shape business process redesign and determine how future information technology will be used. A recursive relationship between business processes and information technology is created. Logically, the more significant the business process is to the organization, the greater the potential benefit from improving that process and its supporting tools that support it. Strategic planning creates the framework for successfully identifying the best candidates for GIS technology application, and for identifying the planning and design process that is most likely to yield the expected benefits. It also provides the framework for ensuring that GIS design aligns with business plans, organizational structure, existing systems, and staff competency. Through participation in the strategic planning process, end users and management build the shared commitment essential for success.

Implementing a GIS can have a profound effect on an organization's structure. The distribution of roles and responsibilities within an organization is to a large extent the result of traditional information-handling tools and procedures (as discussed earlier). Most organizations currently have specialized functional units that concentrate on their own mandates and needs. Effective implementation of a GIS often requires rearrangement of responsibilities across organizational boundaries. The result is a new organizational philosophy, new lines of communication, and a redistribution of business functions. It is impossible to manage effectively the many relationships among people, organizational issues, and technology and across organizational boundaries in the absence of a planning framework.

Several studies have indicated that the scope of benefits is greatest when GIS can be implemented across organizational boundaries. If several organizational units can share data, procedures, and development, both the initial and long-term cost savings are substantial. The Tveitdal and Hesjedal (1989) study identified returns on investment as high as 10:1 when GIS was successfully implemented across organizational boundaries. Achieving success with such sharing requires careful management of institutional issues in concert with technical issues. These arrangements are commonly referred to as multiparticipant projects. This is a complex process made more manageable within the framework of a strategic plan. This subject is discussed in greater detail in the last section of this chapter.

The net result is that benefits and change are tightly interlinked. The purpose of planning is not to eliminate change or to control fully the direction of change, but rather to create a framework within which institutional arrangements and the impacts of change can be effectively managed. Managing change should be a selective process wherein all the positive attributes are accepted and the negative ones are rejected. Strategic planning provides the basis for such decision-making by permitting relationships between cause and effect to be identified prior to implementing change.

There are many approaches to successful strategic planning, and there is no single best approach to conducting strategic planning. The best approach varies from situation to situation. To be effective, each organization or collection of organizations needs to design a process that best matches their particular needs. A well-designed strategic planning process with the appropriate levels of commitment and competence ensures a smooth and orderly transition from today's business processes and systems to tomorrow's. While individual processes are required for each situation, a general framework as presented in Figure 3.4 can guide the planning process. This figure shows that during the planning process a number of things need to happen:

1. An understanding of the organization is developed: its purpose, how it operates, its culture, the management style—*situational analysis*.

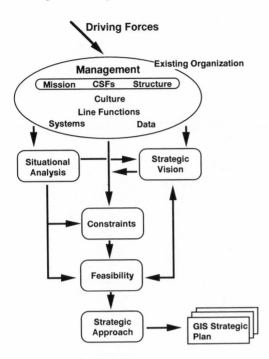

FIGURE 3.4 Strategic planning process.

2. Organizational *constraints* such as available resources (human and financial) and legislative, policy, or management constraints are identified.
3. A vision for GIS use in the organization is established—the *strategic vision*.
4. The strategic vision is aligned with other plans and direction of the organization.
5. Estimates of cost, complexity, and organization impact are prepared in sufficient detail to determine the *feasibility*.
6. The strategic vision is aligned with the constraints, and the feasibility is determined.
7. A *strategic approach* is defined, and strategies for dealing with each of the four aspects of GIS development (concept, data, technology, and organization) are prepared.
8. A *strategic plan* document is prepared as a record.

This does not need to be a lengthy process. The authors have initiated strategic planning processes with intensive 3-day workshops, followed by approximately 2 weeks of followup meetings and documentation.

The second case study described in the Appendix illustrates how this process has been used to create a strategic plan for GIS implementation in the City of

Scarborough. The balance of this chapter explains in greater detail how such a process is implemented.

Monitoring the Plan: Managing Interdependency

As organizations strive to obtain higher levels of benefit from their GIS investments and as GIS applications become more complex, the opportunity for both higher returns and higher failures increases. The investment that the organization makes is larger, and the time frame over which the organizational consequences are felt is longer. Directly and indirectly more of the organization is affected. As the scope of the GIS increases, there is need for more control.

Societal and technological changes are happening very rapidly. Strategic plans can become outdated very quickly. Managers frequently use this as reasoning for not doing strategic planning: Things are happening so quickly; there is an urgency to get on with the job, to concentrate on immediate results without worrying about long-term direction or implications. In this environment, strategic planning seems like wasted effort or an overwhelming challenge. There is a danger of becoming caught in the cycle of reacting to crises and committing no time to planning, thereby ensuring that the crisis cycle continues.

Another view is to see these circumstances as making strategic planning essential. Rather than perceiving the situation as a crisis, management action is brought into alignment with the change. The situation is viewed as an opportunity to tune up virtually every job and function performed by the organization.

Strategic planning should be a process, not a document. It is necessary to have a document that records the agreed-upon strategy, but it should also be possible to make periodic changes as need dictates. Factors internal and external to the organization change constantly. Internally, the organization is developing more understanding and comfort with GIS. Progress may be faster or slower than anticipated. Externally, new advances in technology will have occurred, or the driving forces for the organization may have changed. A good strategic plan provides a framework within which to evaluate opportunities and shortcuts to achieving the vision, and to differentiate legitimate opportunities from unimportant diversions.

The strategic plan should be reviewed periodically, normally annually. A periodic review provides an opportunity for additional alignment of forces in the project, or realignment if some aspects have moved out of scope. Not all aspects of the plan can or should be implemented at once. The project will proceed in predefined increments. One role of the strategic plan should be to coordinate incremental implementation. As the project progresses, the strategic plan provides an overview of the entire project, permitting individual pieces to proceed independently while maintaining the overall direction defined by the strategic vision.

Chapter 4 discusses planning for individual applications and implementation of specific parts of the strategic plan. The implementation plans for individual tasks within the strategic plan must be aligned with each other, with the vision, with the GIS project scope, and with what is feasible for the organization.

Selecting an Approach: Situational Analysis

What is situational analysis, and why begin with it?

Situational analysis is the process of developing an understanding of the present state of the organization or organizations contemplating GIS implementation. It includes analysis of the following:

- *Fundamentals of the organization.* The environment and driving forces motivating the organization, mission, mandate, governing legislation, critical success factors, existing business or strategic plans, goals and objectives, whether the organization is static or dynamic both now and in the foreseeable future
- *Management philosophy and style.* Degree of delegation, position on planning, emphasis on people versus results, past experience with information technology, decision-making style, approach to risk, all of which are usually reflected in the culture of the organization
- *The culture or cultures of the organization.* Mechanistic versus organic; sometimes there are different cultures at the working and the management levels
- *Driving force for GIS.* Who in the organization is pushing for GIS (senior management, a support service such as drafting, middle management, staff professionals) and the impacts of different forces and their effect on the project
- *Technology maturity.* A subjective measure of how much experience staff has with information technology, ranging from never having touched a computer to regular complex use, and evaluation of current GIS usage if any
- *Available resources.* Assessment of available and obtainable financial and human resources
- *The complexity of the business functions of the organization.* A subjective assessment of the type of work performed by the organization without performing a detailed analysis
- *Implementation success risk/uncertainty.* Based on a review of the foregoing, a subjective assessment of the likelihood of successful GIS implementation (adequate return on investment) without major structural changes

Situational analysis provides the conceptual basis for subsequent planning activities. The process reveals the "soft" and "hard" aspects of an organization's makeup. The soft aspects are related to organizational diagnosis such as is performed through process consultation[3]. In its fullness this is a complex process requiring highly trained professionals, often industrial psychologists. It is not the intent to suggest that such thoroughness is required for the purposes of strategic planning situational analysis. Nor is the intent to suggest that implementing a GIS can resolve major organizational development problems. For the purposes of this discussion the intent is to recognize that organizational culture, the relationships between people working in an organization, and management philosophy all influence how people and the organization as a whole will respond to planning, suggested change, and the introduction of new technology.

The hard aspects are structural (budgets, present technology use, organization structure), as would normally be included in a management audit. These reveal practical limitations to what can be accomplished by and for the organization.

In summary, a situational analysis is a nontechnical analysis to determine how the organization will respond to change, how it deals with decision-making, the real and perceived constraints to technology implementation, and the nature of relationships between various operating units and between management and line staff. The underlying premise of situational analysis is that in order to design and implement the GIS effectively and to successfully operate and use it subsequently, the environment into which the technology is being implemented must be understood. A second premise is that the most important factor in successfully implementing and using a GIS is people. People will design and use the system. The correct process well matched to the characteristics of the organization will create an environment of enthusiasm and innovation. The wrong process will at best result in compliance and subdued hostility.

In examining activities that will affect GIS implementation within an organization, the rate and type of change occurring inside the organization can provide many insights into the current situation. Many organizations are currently transforming in response to changing business environments. This change can be characterized as moving from stochastic/segmented organizations to entrepreneurial/integrated organizations. In simple terms, many organizations are changing from static organizational structures based on staff disciplines and business functions, wherein most of the authority resides with management, to dynamic organizational structures based on multidisciplinary project teams that change in response to driving forces in the business environment, and individual project teams are empowered to act as required to achieve the specified results. If this change is occurring in conjunction with GIS implementation, then extra care must be taken to align the GIS planning process with the organizational change strategy. In practical terms it means that more senior management time will be required in determining and executing the GIS planning process. The good news is that management will be predisposed to doing so, or the other organizational changes would not have been initiated in the first place.

A second major indicator of an organization's current situation is its approach towards information technology use.

Information Technology Factors

Different organizations will be at different stages in understanding and adoption of information technology—technology maturity. The planning process should take into account the organization's past experience with computers and its current stage of development with respect to technology planning and management.

The Nolan Model of Computing Technology Evolution

The most widely accepted model of computing growth in organizations was advanced by Nolan, first in 1973 and later refined in 1979. Nolan's (1979) stage model of computing growth explains the growth of computing in an organization as evolutionary, proceeding through six stages:

1. *Initiation.* External technology advancements become known by users and applied to functions requiring cost containment.
2. *Contagion.* Growth in applications increases significantly, and users become more skilled in the technology. Enthusiasm heightens.
3. *Control.* Continued growth causes rising costs, which attracts management attention, and formal planning and controls are implemented.
4. *Integration.* The increase in maturity of computing management identifies the technology as an organization-wide resource, and applications are modified to take advantage of database management capabilities.
5. *Data administration.* The emphasis is on shared data and systems where applications become integrated as a total organizational resource.
6. *Maturity.* Applications are designed to follow the flow of information through the organization, and computing technology is accepted and used as a strategic resource by users as well as managers.

The Nolan model assumes that new technology enters an organization and is adopted at the user level where it becomes popular. It catches management attention when costs escalate, at which point it becomes controlled by top management and eventually matures to the point where it is efficiently and effectively applied to organizational needs.

It would appear that GIS technology implemented in an organization that is in any one of these six stages can experience a different level of success than it would if the organization were in a different stage. For instance, implementation of GIS technology during Nolan's initiation stage might produce applications that are more process oriented, automating high-cost manual drafting tasks in one particular user department, while implementation during the maturity stage might produce the strategic, organization-wide applications envisioned by GIS proponents. If Nolan's work is representative of the evolution of computing in organizations (and it has held up under close scrutiny since), then GIS proponents seeking to initiate a project in an organization would be well advised to determine first the organization's computing stage before presenting the case for GIS. (It is evident that an attempt to sell the benefits of a strategic, organization-wide GIS implementation would fall on deaf ears if the organization were in any of the first three Nolan stages.)

Since different organizations are most likely to be in different Nolan stages, and since they probably progress through these stages at different rates, this model might help explain why the adoption of GIS technology has not been more widespread than expected. It also suggests that understanding who is driving the thrust for GIS implementation is important in selecting a planning approach to implementation.

A Model of Computing Growth in Local Government (URBIS Model)

The most extensive research into local government use of computing technology has been conducted by the Public Policy Research Organization (PPRO) at the University of California, Irvine, and was reported in 1989 (Kraemer et al., 1989). The PPRO began the Urban Information Systems (URBIS) project in 1973 to study local government computerization and identify which policies used for the management of information systems in local government most effectively produced benefits and avoided problems. The URBIS project studied seven U.S. cities and counties longitudinally throughout their entire history (up to 30 years) of computing to identify the forces that affected their "computing package"—a combination of hardware, software, people, skills, operational practices, and organizational expectations—and then to determine the effect of those forces on the use of information systems in those organizations over the years. The results provide some insight into the nature of the environment in which information systems, and GIS in particular, can succeed.

Unlike the Nolan theory of evolutionary stages, the URBIS study found that the manner in which computer technology was controlled within the local governments studied had the greatest impact on how the technology was developed and used. It found that managers in local government have a full range of actions they may take, rational as well as irrational, to influence computing applications in local government. Thus, they conclude, "management matters": "In all seven organizations, of all the factors we studied, the decisions of managers had by far the most significant influence on computing development and use" (Kraemer et al., 1989, p. xii). Kraemer et al. found that the most successful use of computing technology in local government exists when there is congruency between those in management positions who have control over computing policies and those who benefit from them. In other words, success requires that those in control also be those who benefit from computing in local government. Furthermore, they found clear patterns of management and classified them into four different "states of computing management." Each state reflects different managers who can control and use computing technology: the *skill state*, the *service state*, the *strategic state*, and the *mix state*.

The Skill State. In the skill state, the manager of the information systems department controls policies concerning the use of computers and also benefits the most from them. Thus the IS manager is committed to implementing a particular vision of computerization, and the results are an enhancement of the technical quality of the organization. In the skill state, the primary objective of computing activity is to maintain a high degree of technical sophistication with the immediate utility of applications a secondary concern. Other characteristics of organizations in the skill state of computerization include the following:

1. The needs of users are secondary to the opportunity for technical advancement.
2. The applications developed are those of interest to the IS manager.
3. The IS organization promotes expanded technological development.
4. Hardware and software are leading-edge technologies.
5. Hardware is centralized or under central control.
6. Computing staff and budget are large.
7. Applications are often too complicated for many users.
8. Success depends on the vision of the IS manager.

Huxhold (1993) theorizes that if an organization is in the skill state defined by Kraemer et al., then the GIS proponent seeking GIS implementation might expect the resulting system to be under the control of the IS department and to be technically sophisticated, but expensive, complicated for many users, and not linked to the overall strategy of the organization.

The Service State. In the service state, the managers of user departments control computing and benefit the most from the applications. This usually is manifested in the form of a user committee, composed of user department managers who form computing policies for the organization. The resulting objective of computing activity is to support a broad user base with all other organizational concerns secondary. Other characteristics include the following:

1. Hardware and software technologies are diverse.
2. Applications are oriented to support the functions of operating departments.
3. Decisions are made through compromises and coalitions.
4. User charges are subsidized by the general fund.
5. Computing staff and budget are large.

Huxhold (1993) theorizes that GIS proponents in an organization that is in the service state of computing management might expect to seek approval of a user committee that would encourage a variety of different vendor technologies, request applications that are directed towards specific functions, encourage the decentralization of technical responsibilities to the users, and be high in costs and subject to directional changes as interdepartmental coalitions change.

The Strategic State. In the strategic state, top management controls computerization, and the broad interests of top management are served. While top management directs and controls computer acquisition and allocation, the general goals of the organization and the specific goals of top management receive top priority. In the strategic state, the primary objective of computing activity is to support the particular objectives of top management at any given time. Other characteristics include the following:

1. Computing decisions are made by elected and appointed officials.
2. Applications are focused on public services and areas that top management favors.
3. Computing staff and budget are low.
4. User charges help regulate user demand.
5. Agendas, ambitions, and interests of top managers influence computing decisions.

Huxhold (1993) theorizes that in an organization in a strategic state, a GIS proponent should expect the project to be controlled by the mayor, the city manager, or other top elected or appointed official. The system should be expected to reside on a centralized mainframe computer or otherwise be under central control, address uses that are interdepartmental and administrative in nature, and be conservative in technical sophistication and cost. Furthermore, the Kraemer et al. (1989) findings suggest that a GIS project implemented in a strategic state is most likely to succeed because of the organization-wide perspective of the technology, providing that the political interests of top management do not interfere with the long-range goals of the project.

The Mix State. The mix state is a default condition that exists in the absence of any of the three ideal states. The mix state encompasses any set of conditions where the level of control of computing and the level of interests served do not directly correspond. In a mix state, there is no clear direction for computing or the organizational role of computing has not been clearly defined. Characteristics include the following:

1. Any of the three management levels may be in control (data processing management, user management, or top management).
2. Few new applications are implemented.
3. Computing policies conflict or overlap.
4. Computing staff and budgets are large.
5. No power to effect a vision of computing in the organization.

The mix state is an incongruent state, meaning that those in control of the computing responsibility in an organization do not also benefit from its use. This often happens in periods of personnel transition when a city is functioning without clear management direction or when the organizational role of computing has not been clearly defined. Thus Huxhold (1993) theorizes that GIS proponents in an organization such as this might be well advised to postpone their attempt to adopt the technology until one of the three congruent states is achieved, because in the mix state there is no power to effect the vision of computing that GIS proponents have.

The URBIS research can provide GIS proponents with additional insight into strategies for ensuring the successful adoption and implementation of GIS technology in local government. The study found that the key feature of the strategic state of computing is that applications of an organization-wide nature are most likely to be present in this state. Perhaps an organization with its chief elected official (or a designated appointee) making substantive decisions concerning computing technology is most likely to adopt GIS technology because of its organization-wide application. Unfortunately, the findings of the URBIS indicate that the mix state of computing is more likely to be present in local governments than any of the congruent states, which offers a possible explanation as to why GIS technology has not been adopted in local government as widely as would be expected. The future of GIS technology may be optimistic, however, since the URBIS research has observed a trend towards the strategic state.

Both the Nolan model of stages of computing evolution and the URBIS model of management states influencing computing use provide two important considerations for GIS planning and the approach to the project:

1. The management culture and computing environment that organizations are in at any one time differ from organization to organization. This implies that some organizations may be willing to adopt the strategic, organization-wide benefits of GIS technology while others may not.
2. The management culture and computing environment of organizations change over time, so that a given organization may not be willing to adopt the technology at one time but may be willing at another time.

In a strategic sense, then, in order to introduce GIS technology in an organization, it is important to evaluate both its current computing environment and its current management culture with regard to information processing before embarking on a particular path to sell the benefits of the technology. Both the Nolan model and the URBIS model tell us that it is not the technology alone that drives adoption and successful use of GIS technology. This position is further supported by the three case studies presented in Chapter 2.

On a more practical level, it should also be recognized that external technological aspects of GIS implementation are also changing, such as the available technology and its cost, and the methods used to design and implement systems. Planning and design processes have been evolving over the past 20 years. The rapid changes in planning and design methodologies are in large measure in response to the pace with which technology paradigms are changing.

Table 3.1 presents examples of the changes in technology, planning

Table 3.1   Changes to Information Technology

Technology	Systems Design Process	Management Emphasis
Corporate computing. Mainframe with terminals. Many users share one computer.	Specific computing requests designed and implemented by central technical team.	Data center: centralized planning and control. Predefined electronic data processing.
Departmental computing. Minicomputers. Specialized departmental computers.	Paper-intensive specification oriented system development methodologies.	Coordination of distributed implementations. Predefined applications satisfying operational needs.
Personal computing Each person has own computer.	Random depending on the organization. Mostly user-specific, packaged applications.	End-user computing. Isolated personal-use data-bases. Individuals satisfying pent-up demand for access to computer resources.
Network computing. Each person can access many computers and databases.	Corporate data and business models. CASE tools. Diagrams, models, and prototypes.	Central planning and design, distributed processing and data. Business process redesign, end-user management, IT is a corporate resource.

methods, and implementation paradigms over the past 20 or more years. The evolution of information technology paradigms is characterized as four arbitrary but representative eras of information technology use. There is considerable overlap between the eras presented. Several eras may coexist within an organization and certainly can exist for different organizations. The stage of evolution of thinking and implementation becomes another important factor to identify during the situational analysis, and affects the information technology paradigm selected for implementation within specific organizations. This is even more significant when arriving at a consensus among several organizations in a multiparticipant project, as discussed in the last section of this chapter.

Applying the Results of the Situational Analysis

At this point in the planning process, the results of the situational analysis should provide the information necessary to define the best manner in which to proceed with high-level GIS planning. Whether it is referred to as strategic planning or some other process will be a function of the characteristics of the organization and the current state of the organization and its information technology environment. The remaining sections of this chapter present elements of the strategic planning process that have been demonstrated as useful in creating a successful GIS implementation. Which elements are used and the emphasis placed on them will depend on the circumstances, as defined by the results of the situational analysis.

Strategic Vision: Building Commitment

A strategic vision defines the general direction and ambitions for GIS development. The vision should be defined in terms that are meaningful to the organization, using what was determined during the situational analysis as the basis for a vision statement. There are two reasons for developing a strategic vision for GIS implementation: (1) to build commitment for the GIS and (2) to align the direction of GIS implementation with other aspects of the organization.

The content of the statement varies depending on whether GIS projects or related activities are currently under way or if GIS is a new activity. The strategic vision described in Table 3.2 compares the characteristics of a current information technology implementation to characteristics of a future, desired implementation, for an organization that currently has significant amounts of information technology, some of which is used for geographic information handling.

The City of Scarborough developed a different type of vision statement, as illustrated here. While less specific and less quantifiable than the example illustrated in Table 3.2, it clearly states the emphasis on end-user participation in design and operation of the GIS and establishes a quality-oriented goal for the implementation.

> The City of Scarborough will develop the best user-oriented and user-defined municipal GIS in North America.

This commitment to user participation by management and the GIS project team defines end-user commitment as essential to achieving the project objectives. The accompanying process used in Scarborough emphasizes the use of workshops, in-house seminars, a newsletter, and regular presentations to the Council, providing involvement for all levels in the organization and thereby building commitment.

The strategic vision should be a beacon that guides subsequent planning and development rather than a set of specific objectives. The statement should be

Table 3.2   **Example of a Strategic Vision Based on Current and Future Information Technology Implementation**

Present IT Implementation	Desirable IT Implementation
Piecemeal data handling	Continous information flows
Standalone installations	Network of computer systems
Isolated duplicate data handling	Distributed databases: electronic interchange
Low-level technology applications	High-level technology applications
Technical independence	Technical interdependence
Many data models and implementations No geographic data integration	One data model: geographic data fully integrated
Inadequate standards	Standards for data definition and interchange

sufficiently flexible to withstand minor organizational changes and advances in technology over the entire time frame of implementation. While many aspects of the planning process need to be flexible and to evolve as the process proceeds, the vision should remain consistent for the entire implementation. Specific, measurable objectives for individual GIS applications are developed during implementation planning. These can be more flexible from application to application.

The second purpose for developing a strategic vision is to ensure that the goals and objectives developed for the GIS are aligned with other strategies, policies, and related management direction. This is achieved by reviewing existing mission statements, corporate goals, and planning documents, and by involving management in the process of defining the vision.

It should also be recognized that there is a difference between targets and what is actually accomplished at a particular time. The time frame in which the vision is to be achieved can be controlled to account for available resources and acceptable rates of institutional change. Likewise, changes in the cost of technology, available data, and so forth may make today's wild idea reasonable in a year. The next section discusses assessment of the vision's feasibility. Assessing feasibility and establishing a clear vision should be an interactive process during which vision is tempered with reality. Participants should understand the need for realistic calibration of expectations. On the other hand, the vision statement should be bold enough to capture the imagination of participants and management, and to survive changes in technology. The vision should be large enough to encompass future applications in a general way, without being so vague as to provide no direction. It is equally important when establishing a vision to aim neither too low nor too high when establishing a GIS vision.

There are several suggested techniques for arriving at a strategic vision. Some ideas based on observation of many GIS projects in North America are presented in the list following. The projects have ranged in size from small private companies to large government agencies.

1. Involve senior management early in the process. Senior management involvement assigns an appropriate level of importance to the exercise. Strategic planning in many organizations is the domain of senior management, and they must be involved. Additionally, involving senior management in planning and obtaining their commitment through participation will foster support for later aspects of the project. Some organizations have attempted to slip a GIS in under the auspices of something else. At best, this approach will lead to a tolerated project that limps along without adequate resources. At worst, it will result in absolute failure, wasted resources, and exclusion of future technology.

2. Involve end users in a similar fashion to senior management. Bring them in at the beginning. Keep them involved throughout the process. Not all staff can be fully involved in doing the work, but everyone can be informed that the activity is in progress. Produce a letter from the head of the organization, or have a general meeting to inform people of what is going on.

3. Appoint a task force or working group to facilitate creation of the vision. Use workshops to involve others to the extent required. Create a brief written statement of the vision and circulate it to management and staff. Prospective users may not be familiar enough with GIS to know what to expect. Process leaders should be prepared to use seminars, demonstrations, and prototypes to raise the awareness of potential users.

4. Assemble existing documents, mission statements, goals, and other documents to summarize the key issues and beliefs of the organization.

5. Use workshops to solicit input and to arrive at consensus. This can be done in conjunction with defining the planning process. If there are contentious issues to be resolved or no one feels comfortable leading the workshop, bring an outside facilitator to lead the discussion.

As stated earlier, this need not be a lengthy complex process. A 1- or 2-day workshop, followed by a week of small group discussion and documentation, will usually suffice, even in large organizations.

Determining Feasibility: Establishing Project Scope

So far, this discussion of strategic planning has emphasized building alignment with the organization and its business functions and fostering commitment from the various participating parts of the organization. In this section, the focus is on bringing the vision in line with reality. Two topics will be addressed together: determining feasibility[4] and establishing project scope. During the strategic planning phase, both of these topics are dealt with at a general organization-wide level. Each topic will require further investigation in subsequent phases of GIS implementation.

Feasibility is measured in terms of financial, technical, and institutional factors. The manner in which these are measured depends on the organization. As discussed in the section on situational analysis, attitudes towards risk, requirements for measurable decision parameters, and the planning horizon of the organization determine how feasibility is measured. However, the items in Table 3.3 are representative of what should be evaluated while determining financial, technical, and institutional feasibility.

Of initial importance is establishing the planning horizon over which feasibility will be determined. A short planning horizon of a year or less will cause all but the simplest GIS applications not to be feasible. Similarly, a 10-year planning horizon may make everything feasible in theory, but reliable measures to support the findings cannot be established. A reasonable planning horizon is 3 years. Most internal and external changes and trends can be adequately predicted for a 3-year period. Additionally, the planning process can be structured to be more detailed for the first year and less so for the following 2 years. An annual update to the plan can incorporate new information as it becomes available. A grand vision may require a longer planning horizon, but

Table 3.3   Items to Be Included in Evaluating Feasibility

Financial Feasibility

- The anticipated cost of GIS development is within the realm of resources that can be expected during the planning horizon
- The returns to the organization warrant the planned expenditure
- Adequate performance measures can be determined to satisfy the organization's management and policy requirements

Technical Feasibility

- The required technology is currently available or will be available within the planning horizon
- The level of technology required to computerize identified functions is reasonable and practical
- The technology required can be made usable by intended staff with reasonable amounts of training

Institutional Feasibility

- An adequate level of funding can be sustained over the planning horizon
- The task competency required for GIS development design and management is available or obtainable
- The anticipated levels of change to the organization's structure and business processes are acceptable
- The proposed implementation and the impacts thereof do not violate legislation or mandates governing the organization

---

specific elements of the vision should fall within the 3-year planning period, and be therefore definable.

Financial Feasibility

Financial feasibility should be determined throughout the phases of GIS implementation. A general investigation of costs and returns to the organization may be adequate at this stage, followed by more detailed analysis as part of the process of selecting and implementing specific GIS applications.

In discussing financial feasibility it is necessary to begin with a brief discussion of cost–benefit analysis. Cost–benefit analysis is the recognized approach to choosing between alternative courses of action in business practice. Costs are basically grouped into capital, one-time, and ongoing. Benefits can be broadly grouped into cost savings and cost avoidance. Many organizations have used the technique to justify expenditures for GIS implementation. In many organizations a cost–benefit study is required by policy or legislation.

There have been countless books written on cost–benefit techniques. There have been even more articles written favoring or rebuking cost–benefit analysis as a basis for deciding on information technology investment. In the end, whether and to what extent a cost–benefit analysis is conducted will depend on the policies and practices of the organization. A mechanistic organization accustomed to dealing with hard facts will be more likely to demand a cost–benefit analysis than a more organic organization that will be satisfied with "softer" analysis of costs and benefits. Organizations such as government, which spend tax dollars, may be required by legislation to conduct a cost–benefit analysis for all major expenditures.

Not withstanding the foregoing, there are some reasons for questioning the relevance of traditional cost–benefit analysis for information technology implementation:

1. The costs of the undertaking are normally easily collected, but many of the benefits are intractable because often the most important benefits are intangible or unquantifiable (Gotlieb, 1985).
2. While it may be possible with adequate persistence to translate intangible benefits into measurable parameters, it also adds significantly to the cost of conducting a cost–benefit study.
3. It is difficult (or impossible) to measure the benefits (or impacts) to an organization for a new process that will change the functioning of an organization, particularly when there is a high likelihood that existing functions will be redistributed within and across organizations.
4. There is a growing body of evidence to suggest that cost–benefit analysis is biased by the need to establish the investment criteria (e.g., acceptable return on investment) at the outset, thereby creating an invitation to bias the analysis and making the process more political than analytical (Goodman and Sproull, 1990).

Perhaps the most significant reason to question the worth of traditional cost–benefit analysis for GIS is that in the first instance the decision is really whether to automate or not to automate. If we assume that the organization is not fundamentally opposed to any expenditure on information technology, the question then becomes how much the organization is willing or able to spend and over what time period. The question is therefore not one of determining the costs and benefits of GIS, but rather one of choosing an optimal level of expenditure. An optimal level of expenditure is closely aligned with the strategic vision and the planning horizon selected. Often ongoing implementation and use of GIS technology is financed by the redirection of operating funds. As various processes are computerized, the resultant savings are directed toward additional GIS technology implementation.

The strategic vision should define the desired level of computerization. That vision must now be adjusted by what is financially reasonable for the organization. The allowed, affordable, annual expenditure helps to establish both project scope and the planning horizon over which implementation will occur. A grand vision can be achieved with a small annual budget by extending implementation over a longer period of time. There are, naturally, practical limitations to how little can be spent while still producing a meaningful project.

These points not withstanding, it is necessary to identify accurate estimates of the costs over the planning horizon and to ensure that the most appropriate expenditure for technology is being made for the most appropriate applications. The question is, then, if a formal cost–benefit analysis is not performed, how can this be accomplished?

Cost of implementing and operating can be weighed against the costs of manual title searches, manual map production, couriers, searching for information, not having the information, incomplete records, and damage to public image and relations resulting from errors. While these may be difficult to quantify in great detail, the aggregated budgets for these activities are usually available. It is also reasonable to assume that all costs associated with manual procedures will increase in future, while many of those associated with automated procedures will decline.

Consideration should be also be given to the fact that information is subject to increasing returns with use when properly managed and accessible, because it gets less expensive per use after initial collection. In general, the major costs of implementing a GIS are data related: collection, conversion to digital format, database design and implementation, and ongoing maintenance. Estimates range from 65 to 80 percent of total implementation cost. Therefore, design strategies and organizational approaches should stress the data and the procedures associated with data maintenance and dissemination. Hardware and software are exchanged as more cost-effective technology emerges. Data, on the other hand, can be perpetuated if properly managed. Given the foregoing, improvements to safe guarding, usability, and maintenance of data should be given prime consideration in cost-justifying GIS implementation.

If a formal investigation of cost–benefit is required, consideration should be given to hiring an expert in these studies. First, these studies can be complex, and cost justification is not always immediately apparent, and second, an organization that requires a comprehensive cost–benefit analysis will not likely trust the results of an internal study.

Technical Feasibility

Technical feasibility is an equally complex issue, and the measure of feasibility depends on many factors. Of greatest significance is having criteria by which to select the most appropriate technology for each application or set of applications. These criteria are the result of a detailed understanding of the business functions to be supported and to what level of sophistication. Basic support functions such as data query and display are far less complicated to implement than computerizing aspects of complex decision-making.

Without adequate decision criteria directed by the strategic vision, there is a strong likelihood of becoming "technology driven." That is, new developments in technology are justified on their own merit. Solutions go looking for problems to solve. In this environment technical feasibility ceases to be an issue. The right technology is always available, because "right" is defined by what is available. A broader more systematic view of technical feasibility is required to avoid becoming technology driven.

Some significant factors from the perspective of overall technical feasibility are

- Balancing complexity with competency and need
- Providing different levels of technology to different users depending on need
- Finding ways to provide affordable technical support when the need demands a level of technical complexity greater than the end user can support
- Resolving the technical problems that result from acquiring technology over a long period of time, including managing hybrid technology and migrating data and applications to new technology

In evaluating technical feasibility, part of the evaluation relates to use by the end user and to the amount of training and technical support necessary to use the application. One way of dealing with technical complexity is to hide it from the end user by implementing sophisticated, easy-to-use machine interfaces. The human–machine boundary is a conceptual meeting point of the operator and the computer. The more of the complexity of the workings of the computer that can be hidden from the user, the easier the computer is to use. More "user friendliness" generally means more software development, but also lower training and startup costs for the end user. This concept is well characterized by the difference in user interfaces between the Apple Macintosh and the IBM PC (before Microsoft Windows 3.0).

It has been demonstrated through various projects that significant reductions in training requirements and learning curves can be achieved by customizing software for specific job functions and through effective user interfaces. However, that often translates into higher development costs per application unless the software development can be shared. Sharing implies some degree of standardization of design approach and implementation. Standards, in turn, depend on a willingness to cooperate and the sacrifice of some autonomy. Such arrangements can best be dealt with during the strategic planning process and established as implementation principles, rather than attempting to deal with issues individually as applications are implemented. As a final note on user interfaces, rapid advances in standard user interfaces (e.g., Motif, Windows 3.0) to GIS software are reducing development costs, and the issue of development cost will likely be of less significance in the near future.

Technical feasibility is further complicated by the likelihood that GIS technology will be acquired over extended periods. Acquiring technology over a period of time has both benefits and risks. By purchasing only what is required for each application, the organization can benefit from the rapid advances in information technology. The risk is not being able properly or readily to combine the technology. The risk can be managed by defining a technology strategy and architecture, by adhering to as many industry standards as practical, and by closely monitoring technology trends. The process and issues in defining a technology architecture are dealt with in Chapter 4.

Institutional Feasibility

Institutional feasibility deals with the willingness and ability of the organization to accommodate change and to work across traditional lines of authority. When a planned GIS implementation will serve many organizational units and require several years to complete, institutional feasibility is as important as any other. Financial and technical feasibility need to be aligned with the ability and willingness of the organization to sustain a large project over the life of the planning horizon. Budgets and management support will need to be sustained at required levels, staff education and training may need to span several years, and technology acquisition may be spread over several fiscal years.

Institutional feasibility can be assessed by looking at administrative policies and budgeting and planning processes. However, it can best be assessed through discussion with senior management to determine their willingness to commit to longer-term projects. If the organization's policies do not readily allow for one longer-term project, then several smaller related projects may be more appropriate. If management does not accept integration of systems and processes across organizational boundaries, then a strategy that relies upon multiple agencies cooperating in the development of the GIS will not work. Less direct mechanisms such as coordinating committees may suffice or in final analysis the project as conceived may not be feasible.

Institutional feasibility, more than technical or financial feasibility, is tightly bound with the scope of the project. The results of the situational analysis will have provided most of the information required to determine the strategy for proceeding and suggest a project scope. Some additional factors to consider in evaluating institutional feasibility and establishing the project scope are discussed in the following:

1. Is the GIS to be multiple or single purpose?

A system designed to support a single organizational function is simpler to specify, design, and implement than one that must support a variety of functions. However, if the single-purpose implementation depends on data from other parts of the organization, or if other parts of the organization also have an interest in GIS, then a single-purpose project attempt may fail. Implementing a GIS in one organizational unit and not in other related units may also create imbalances in the overall functioning of the organization.

2. Will the GIS be implemented to automate line or support functions?

Many organizations have begun by implementing computer-aided drafting systems to automate map production. That technology will likely not be adequate to support line functions that require complex information retrieval and modeling. If the technology is to be limited to support functions, that must be clearly understood by users to prevent unrealistic expectations.

3. For whom is the computer system being developed, and what type of computer system is required?

If the system is being implemented to support management planning and decision-making, the system will be significantly more complicated than if it is being implemented to perform routine information handling tasks. Martin classifies information technology into the following categories:

- Routine processing systems consisting of predefined tasks, and large volume throughput, making up the bulk of conventional data processing
- Information systems dealing with unstructured, unanticipated queries, or queries that trigger subsequent operations, the main function of which is to produce reports
- Decision support systems dealing entirely with unstructured use and unanticipated queries, designed to support executive decision-making
- Operational research or special computation systems, designed to support specific research computation (Martin, 1982)

This classification is general to all computer systems, but applies particularly well to geographic information technology. Design and implementation vary significantly among the different classes of system, and it is important to determine which system is being developed prior to beginning the design process.

4. How integrated will the new system be with organizational functions?

Further to the foregoing, the degree to which the resultant system is to be integrated into the daily workings of the organization must be established. A system that will be isolated from line staff and be supported by technology experts will have different design characteristics, cost, and institutional impact from a system that is routinely used by line staff.

Many organizations do not adequately define the scope of the system to be developed before beginning implementation. This results in systems that do not meet expectations because developers and end users have different impressions of what is being developed, or significant cost and time overruns occur as more capabilities are added to the system to satisfy end users, or combinations of the two.

The scope of the project should be clearly defined and be the subject of workshops or meetings with the key participants in the project. The scope should be a more concrete expression of the vision statement developed earlier in the process. The scope must be aligned with financial, technical, and institutional feasibility to ensure that expectations can realistically be met.

Strategic Planning for Multiparticipant Projects

The preceding sections have largely focused on GIS implementation within a single organization or a group of operating units within the same organization. However, increasingly, projects involve more than one organization or multiple diverse agencies within a large organization. In both cases GIS implementation crosses established organizational boundaries, creating additional technical and management challenges. In this section we apply the ideas and techniques from earlier sections to multiparticipant GIS implementation—projects in which several different organizations are participants. In a multiparticipant project, the organizations may vary substantially in mission, management philosophy, culture, and stage of technology use. The planning and specification of these projects present a special set of issues from those encountered in single organization projects.

Perhaps the first question to ask is, when is a project a multiparticipant project and/or when is it advisable to organize a GIS project as a multiparticipant project? We will therefore begin by investigating some of the more significant characteristics of a multiparticipant project, followed by a review of structuring and managing multiparticipant projects. The specifics of conducting a multiparticipant project will be covered in Chapter 4, dealing with implementation planning.

What Is a Multiparticipant Project?

Many projects are multiparticipant projects without the participants recognizing it at the outset. In general, any GIS project that involves more than one user, each of whom has a different reason for implementing a GIS, is a multiparticipant project. In practice, the difference between multiparticipant and single-participant projects may be measured in the degree to which the differences are recognized and addressed during implementation. Many projects in large organizations are multiparticipant projects. They may not be recognized and managed as such, but they involve more than one department, each with their own business functions, data requirements, and reasons for wanting to implement a GIS. In some instances the failure of GIS implementation may be attributed to an organization's failure to recognize a GIS project as a multiparticipant project. Successful implementation of projects involving more than one set of stakeholders most often requires recognizing and successfully dealing with the needs of all participants, individually and collectively.

There can be many forms of multiparticipant projects and many reasons for the creation of a multiparticipant project. Projects may involve more than one level of government, private companies such as utilities, or private investors in some form of joint venture. Many groups of diverse stakeholders are common in state and province-wide GIS coordination projects. Projects can be short term to meet specific objectives, terminating when the objectives have been met, or long term permanent arrangements. They can be highly structured, involving legal agreements, or more loosely structured, coordinating bodies focusing on standards and cooperation among separate implementation projects.

Many multiparticipant projects result from recognition of a need to share some aspect of GIS implementation—data, funding, development, coordinated access to clients, technology. In these instances, the creation of the multiparticipant project is driven by interdependent needs, such as a desire to benefit from sharing and the result of relying upon each other for data, technical support, and/or money. Realizing the mutual benefits from cooperation may be the result of careful analysis and planning or just running into brick wall with individual projects.

How Do GIS Multiparticipant Projects Differ from Other Multiparticipant Information Systems Projects?

Many systems development projects now span several organizations, and one integrated system often satisfies the needs of several user groups. Many organizations are also reorganizing business processes based on the new systems integration. However, while GIS projects may include many of the characteristics of other systems integration projects, they are a special case because of the focus on geography as the common denominator for data and business process integration. Few organizations link their business processes on the basis of geography. Traditional grouping of business functions follows the purpose of those functions and frequently by the education required to conduct them. Hence we separate engineering and planning departments, oil and gas exploration is separate from lands administration, and so forth. Implementation of a GIS will link these previously disparate business activities and databases using a shared geographic database as the unifying denominator.

How Are Multiparticipant Projects Organized?

There are many project characteristics that define the structure of a multiparticipant project and many different forces driving the creation of one. Figure 3.5 shows some of the common characteristics that define the structure of multiparticipant projects. Each characteristic is a continuum from wholly one orientation to the other. Different projects will be characterized by different combinations of characteristics. Additionally, projects may be designed to share any or all of the data, technology, funding, and development. Which characteristics of project structure apply and to what extent and which aspects of the GIS implementation are shared are defined by the characteristics of the participating organizations and the forces driving the multiparticipant relationship. For example, declining government budgets have contributed significantly to the cooperative efforts to share data among government agencies.

The main point is that there is no single successful organizational structure for multiparticipant projects. As with GIS implementation in single organizations, discussed earlier, the combination of organizations and the project objectives define the characteristics of the project structure. Successful multiparticipant project organization depends most on successfully aligning the characteristics of the project structure to those of the participating agencies, individually and collectively. This subsequently provides a solid basis for defining which aspects of GIS implementation will be shared and how.

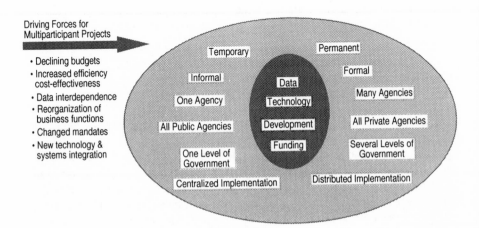

FIGURE 3.5 Characteristics of multiparticipant projects.

The underlying objective is to establish a structure that minimizes the potentially adversarial environment created in bringing multiple participants together, each of whom has a stake in the implementation of the GIS, but for a different set of reasons, without overly complicating the project or losing sight of the original objectives. Figures 3.6 and 3.7 present two examples of multiparticipant projects and the characteristics of organizational structure used in each case.

The Alberta Land-Related Information Systems Project, as shown in Figure 3.6, is representative of a very large multiparticipant project. This structure of committees was used during the formative years of the project and represents the need to involve and coordinate many government agencies and interest groups. The project is operational now, and the complex committee structure has been replaced with a more streamlined project structure, but many of the advisory committees providing user input still exist. This sort of structure has also been

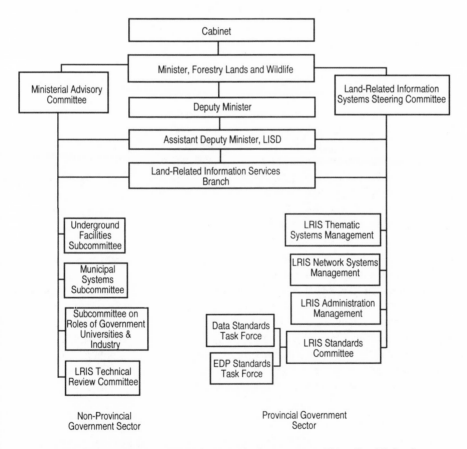

FIGURE 3.6 Multiparticipant of GIS project organization structure: Alberta Land Related Information Systems Project Committee example.

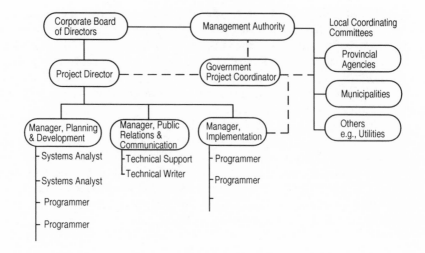

FIGURE 3.7 Multiparticipant GIS project organizational structure: government and private-sector consortium.

used in other jurisdictions, such as the State of Wisconsin Land Records Modernization Project.

In the example shown in Figure 3.7, the project members are a private-sector consortium and the government. The GIS implementation is being undertaken as a private venture under contract to the government. The private consortium will contribute toward the cost of development in exchange for ownership of the developed technology and implementation support from government staff. The management authority provides for the overall coordination of the project and deals with the private consortium. Local coordinating committees provide stakeholder input to the project and access to staff and information required for implementation. Detailed technical and implementation coordination is achieved by the private-consortium analysts who must resolve technical issues to deliver a successful project.

Although project structures will vary from project to project as a result of the issues discussed, there are some common characteristics that are shared by most multiparticipant projects. In almost all multiparticipant projects there are underlying reasons driving the multiparticipant process, there is most often a lead agency that promotes participation by others, and there is some form of committee structure that facilitates communication and coordination. These form the basis for structuring a multiparticipant project.

A Motivation for Working Together

All human action is rooted in some form of motivation. Likewise, participants in a multiparticipant project are for some reason motivated to work with each other. The motivation may be shared or each participant may have his or her own reason for participating, or more likely the reason is some combination of the two. Some typical reasons why agencies work together are: being directed to so; declining budgets and therefore a need to become more cost effective; opportunities for data sharing and more effective access to data; revenue generation through sale of data and services; information systems redesign and integration; and reduced costs and increased profits.

That individual agencies have their own objectives is not a problem unless the objectives conflict with each other. One of the purposes of the situational analysis is to determine potential conflicts and to identify means to resolve them.

A Lead Agency

Often there is a lead agency who by mandate, need, or self-interest will initiate a multiparticipant project. Often these agencies have traditionally had a mandate for land and resource information collection, mapping, aerial survey, planning, or legal survey. Other times the lead agency may be a specially appointed senior-level task force, a policy board with broad mandates, or an organization such as a planning committee that traditionally has facilitated coordination of various issues. In government, individuals within these organizations have often taken the lead in encouraging the development of a shared GIS as part of a program to increase the effectiveness and efficiency of government agencies.

Within private companies, multiparticipant projects are often identified during business process redesign projects focusing on reducing operating costs. Intrapreneurship and competition between companies has fostered innovation in GIS implementation. Corporate planning groups and special task forces review company operations with the objective of streamlining operations and determining new working arrangements between departments. Such groups under the direction of senior management create multiparticipant projects to redefine business processes and implement GIS technology to support the redefined business processes.

A Committee Structure

There is almost always a committee structure of sorts. It may be complex, with several levels of committees coordinating the activities of many agencies at several management and working levels, or it may be relatively simple, consisting of a single committee that both coordinates and performs the required work. Committees often bear other designations that reflect the culture of the organizations: working group, task force, analysis team and so forth.

A multiparticipant project should almost always report to a policy-level committee. A new committee can be created specifically for the project, or an established committee can have the project as a regular agenda item. The latter is usually preferable due to crowded schedules of most senior management.

However, for large projects or instances where the different agencies did not previously interact on a regular basis, a new committee may be necessary.

Reporting to the policy-level committee should then be one or more working groups. Again the number and structure of these groups depends on the size and complexity of the project. Much of the work and detailed coordination may be achieved with contracted staff or a venture with a private firm may formed, as in the example in Figure 3.7. A permanent secretariat may be created to satisfy the day-to-day administrative needs of the project. This is generally more successful than assigning project staff to any of the participating agencies because of the potential for mixed loyalties and conflicting management direction.

The resultant project structure should not be more complicated than necessary. Too elaborate a project structure for a straightforward project is as likely to result in failure as too simple a structure would for a complex project. If the project participants have close working relationships, high levels of trust exist between participants, and the project objectives are relatively simple (e.g., cooperation in the development of data exchange standards). In this case a simple forum for discussion and a document exchange structure will likely suffice.

On the other hand, if the land and resource activities of an entire state or provincial government are being coordinated, as in the Alberta example given, then a complex structure is necessary to provide mechanisms for involvement for all stakeholders. Project structures can evolve over time, becoming more complex as need warrants. Similarly, the project planning should provide for dissolution of committees and working groups as their work is completed.

Planning Multiparticipant Projects

In multiparticipant projects, more than in single-agency projects, it is important for a planning framework to be created at the beginning of the project. With many agencies or departments involved, managing the many required activities and keeping participants informed of progress becomes very challenging. It is also important for the various participants to support the implementation approach and process selected through participation. To do so, the process must be adequately defined and communicated to all participants early in the project. A well-designed planning and implementation process establishes a balance between the details (microconcerns) of each agency and implementation task and the broader longer-term implications (macroconcerns) of GIS implementation and business processes redesign. Adequate communication during implementation can help prevent one of the problems identified in the USAC project described in Chapter 2. It helps to ensure that all participants share the common goal that reflects the individual goals of each participant.

The planning activities for a multiparticipant project are much the same as those discussed earlier in this chapter, except that the process must be conducted for each of the participants individually and all the participants collectively.

Present and planned operations must be defined, understood, and agreed upon. The present organizational structure, technology, and operations within each of the agencies must, of course, be determined. However, equally important and more difficult, the new organizational structures, technology, and operations following implementation of the shared GIS implementation must be estimated. Each agency will want to understand the impact shared GIS implementation will have on its present operations, organizational structure, and budgets, as well as what will be required of the agency in future under the new operating arrangements. Figure 3.8 presents the process.

The starting point is a thorough situational analysis that identifies the characteristics and needs of each participant. It should identify differences in the operating policies, management philosophy, and organizational culture among participants. This is important for subsequently defining the institutional arrangements among participants. Practical issues such as available funding, information technology policies and current investment, skill levels of staff, and past experiences with information technology are all important in later determination of technology strategies and architectures.

The results of the situational analysis in a multiparticipant project also help to define the best approach to proceeding with planning. GIS implementation is often slow, especially when significant databases need to be constructed prior to developing any applications. Tangible results are often slow in materializing, especially for agencies such as planning and resource analysis where large amounts of diverse data and higher skill levels are required. The participating agencies will be more likely to be understanding of delays and interim implementation activities if they participated in determining the process used to define, design, and implement the GIS. Additionally, they will be concerned

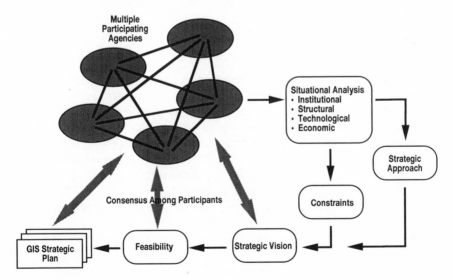

FIGURE 3.8 Multiparticipant GIS strategic planing process.

with ensuring that their interests are being well served. By having participants agree to the process and objectives of the project, a common basis for proceeding can be defined, even when tangible results are not immediately forthcoming.

The key issues of each participant and any major institutional impediments will be identified. Identifying and understanding the issues and institutional concerns among participants helps to develop a consensus in later stages of planning and project implementation.

The situational analysis is often performed by the lead agency in preparation for identifying the appropriate stakeholders and their subsequent roles in the project. Once a planning framework and goals for the multiparticipant project have been defined, priorities for specific activities can be defined. A conceptual model of the GIS can be developed and application and data management specific implementation tasks can be defined.

In conjunction with the planning activities, the institutional arrangements between participants need to be determined. Issues that need to be resolved include: who will be responsible for various aspects of GIS implementation and subsequent operation; cost sharing; data access arrangements and costs; and reorganization of business processes, as necessary. It is these high-level issues that require involvement by senior management, and thus the need for the project to report to a policy-level committee. Multiparticipant projects not fostered through policy-level support often fail, or stagnate for long periods of time while policy-level support is acquired to resolve major issues.

As has been stressed in the preceding discussion, there are many project structures that will result in successful multiparticipant projects. The correct structure depends on the aforesaid characteristics. Presented in the following are two examples of project structure characteristics and when they might apply.

Temporary versus Permanent

Multiparticipant projects can be permanent[5] or temporary. Permanent multiparticipant projects are usually established for large-jurisdiction or corporate-wide projects. In a permanent project there is an ongoing requirement to coordinate the business affairs of participating organizations. Within governments these projects are often established as new organizational entities with mandates to develop and implement a multiparticipant GIS across government departments. Multiyear budgets and new operating policies are often developed to facilitate project implementation. The end result of the project is often a redefinition of government programs and budgets. Many such initiatives have been in progress for 15 or more years.

A temporary project is usually single purpose, and once the purpose has been achieved the project is disbanded or reconstituted for new purposes. Temporary projects often take the form of task forces or working committees under the auspices of an established body.

Coordination versus Joint Venture

Another defining characteristic is whether the project is intended to provide coordination between existing organizations and programs or whether a new entity will be created. Related to these are whether existing programs will continue largely as is with minor adjustments to accommodate the needs of other agencies, or whether the objective is the replacement of all or portions of existing programs. These subsequently lead to a number of other issues related to funding, reallocation of responsibility, and potentially to changes to legislation, in the case of government implementations.

In a coordination project, the emphasis is on coordinating separate projects, each leading to the development of separate systems, leaving existing projects intact. There are mutual interests such as sharing data that provide the motivation to cooperate. These projects may or may not include some shared funding, but usually do not result in changes in budget processes. They are usually less formal in structure and may not have clearly defined deliverables. They typically consist of a committee or other communication mechanism whose principal function is to share information about what each participating agency is doing. In government projects, if private companies are participants, it is normally as advisors or observers, not as equal participants. Private multiparticipant projects are almost always joint ventures. Other less formal requirements are normally handled through existing industry associations.

In a joint venture there is typically a specific purpose such as the creation of a new jurisdiction-wide database or map series that cannot be funded by a single agency or company. Joint ventures may also be set up to manage and fund development and implementation of new technology frequently associated with outsourcing[6] arrangements. Joint ventures are more formal and almost always require cash contributions from each participant towards achieving the objectives of the project. Joint ventures can consist of only government agencies or private companies or a combination of government agencies and companies. Joint ventures between different levels of government and private companies are becoming more popular. In such ventures government often provides access to the data, expertise in land-related business processes, and some of the required funding. Companies will provide technology expertise, information technology, and the remaining investment.

Projects can be set up to deal with any or all of the aspects of project development: funding, technology, data, applications design. Some recent examples include a project to develop a shared cadastral and utilities data model by a group of Ontario municipalities, privatization of land registry data handling functions, and the funding of base mapping.

Joint ventures are out of necessity more formal business arrangements. Contracts and/or memoranda of understanding are signed by the participants. These documents clearly state the responsibilities of each party in the agreement and the available recourse if one or more of the parties should default on the terms, just as in any other legal agreement.

Although joint ventures can be more complex and costly to establish, they also tend to deliver more and greater value results for the participants. As GIS technology becomes more complex and expensive to implement, and more successful joint ventures become known, it is likely that such arrangements will become more popular.

NOTES

1. The *Oxford English Dictionary* defines a *paradigm* as a pattern or example. The term *paradigm shift* is used to describe a change in the broad model, framework, way of thinking, or scheme for understanding reality: changing from one model to another.

2. Edward de Bono has held faculty appointments at Oxford, Harvard, Cambridge, and the University of London. The founder and director of the Cognitive Research Trust, he developed the concept of *lateral thinking*. Most of his work is devoted to the elaboration of often radical approaches to creative thought.

3. Schein (1969) defines process consultation as "a set of activities on the part of the consultant which help the client to perceive, understand, and act upon process events which occurs in the client's environment." The process consultant focuses on the assumption that a client's real problem is often that he does not, nor should be expected to, know what the problem is. The process consultant seeks to give the client insight into what is going on around him, within him, and between him and other people. The concept is based on the reality that the interactions between people must be taken into account when developing solutions to technical problems (pp. 3–9).

4. A solution is *feasible* when the values of certain variables satisfy constraints imposed by the nature of the problem. Examples are

- Mix problems in linear programming, where the amount of an ingredient cannot exceed a given available supply
- Class–teacher scheduling problems, where each teacher can meet with at most one class, and each class can meet with at most one teacher in a given hour; this constraint is imposed by the logic of the problem. There may be others introduced arbitrarily, for example, prescribed maxima and minima for the number of sessions in a week for any teacher
- Routing problems, as in scheduling commercial aircraft, where the number of hours a plane may fly before it returns to its home base for maintenance is limited by regulation

In each case the technique is to find, among the feasible solutions, the one (or perhaps the set) that satisfies some condition of optimality (e.g., least cost or fewest personnel required). In computer feasibility studies it is usually expected that factors other than cost will also be taken into consideration. Overall system performance is examined, including such features as throughput, response time, reliability, availability and quality of software, vendor support, possibilities of subsequent upgrades, and so on (Gotlieb, 1985, pp. 125–26).

5. In this instance, *permanent* is used in a relative sense referring to a structure that has been established for some ongoing purpose, rather than a single project of fixed duration.

6. Outsourcing is a contractual arrangement in which an external company takes over the information systems operation, delivery of services, and sometimes planning of an organization. See also Chapter 6.

REFERENCES

Brown, G. E., Jr. (1991). California congressman has GIS vision. Keynote address, GISDEX Conference, Washington, D.C., quoted in *GIS World,* September.

Campbell, H. (1992). Organizational issues and the implementation of GIS in Massachusetts and Vermont: Some lessons for the United Kingdom. In *Environment and Planning B: Planning and Design,* vol. 19, pp. 85–95. Pion, London.

Computing Canada. (1992). *GIS Growth Predicted,* vol. 18, no. 1, p. 23.

Croswell, Peter L. (1989). Facing reality in GIS implementation: Lessons learned and obstacles to be overcome. In *Proceedings Urban and Regional Information Systems Association Symposium,* Boston, 1989, vol. 4, pp. 15–35.

DeMarco, Tom, and Timothy Lister (1987). *Peopleware: Productive Projects and Teams.* Dorset House, New York.

Goodman, Paul S., and Lee S. Sproull (1990). *Technology and Organizations.* Jossey-Bass, San Francisco.

Gotlieb, C. C. (1985). *The Economics of Computers: Costs, Benefits, Policies, and Strategies.* Prentice-Hall, Englewood Cliffs, N.J.

Hughes, John (1990). GIS job opportunities abound: Salaries good but highly variable. *GIS World,* December.

Huxhold, W. (1993). The application of research and development from the information systems field to GIS implementation in local government: Some theories on successful adoption and use of GIS technology. In *Diffusion and Use of Geographic Information Technologies,* Ian Masser and Harlan J. Onsrud (eds.), chap 1.4. Kluwer, Deventer.

Jones, C. (1981). *Programmer Productivity: Issues for the Eighties.* New York Institute of Electrical and Electronic Engineers, New York.

Kraemer, K., J. King, D. Dunkle, and J. Lane (1989). *Managing Information Systems: Change and Control in Organizational Computing.* Jossey-Bass, San Francisco.

Martin, J. (1982). *Strategic Data Planning.* Prentice-Hall, Englewood Cliffs, N.J.

Nolan, R. (1979). Managing the crisis in data processing. *Harvard Business Review,* March–April.

Schein, Edgar H. (1969). *Process Consultation: Its Role in Organization Development.* Addison-Wesley, Reading, Mass.

Scott, W. R. (1987). *Organizations: Rational, Natural, and Open Systems,* 2nd ed. Prentice-Hall, Englewood Cliffs, N.J.

Tveitdal, Svein, and Olav Hesjedal (1989). GIS in the Nordic countries: Market and technology strategy for implementation—A Nordic approach application of GIS to forestry and natural resources—Norway. In *Proceedings from the 1989 GIS Symposium,* Vancouver Canada, March 7–10, 1989, pp. 201–11.

Walton, Richard E. (1989). *Up and Running: Integrating Information Technology and the Organization.* Harvard Business School Press, Boston, Mass.

ADDITIONAL READINGS

Booth, G. M. (1983). *The Design of Complex Information Systems: Common Sense Methods for Success.* McGraw-Hill, New York.

Croswell, Peter L., and Stephen R. Clark, (1988) Trends in automated mapping and geographic information system hardware. *Photogrammetric Engineering and Remote Sensing* 54, no. 11:1571–76.

Fisher, Peter F., and Richard E. Lindenberg (1989). On distinctions among cartography, remote sensing, and geographic information systems. *Photogrammetric Engineering and Remote Sensing* 55, no. 10:1431–34.

Foley, Maurice E. (1988). Beyond the bits, bytes and black boxes. *Proceedings GIS/LIS 1988,* San Antonio, vol. 2, pp. 608–17.

Forrester, Jay (1965) A new corporate design. *Industrial Management Review* 7, no. 1.

Kraemer, K., and J. King (1979). A requiem for USAC. The Regents of the University of California, Irvine, California.

Martin, J., and C. McClure (1985). *Structured Techniques for Computing.* Prentice-Hall, Englewood Cliffs, N.J.

Yourdon, E. (1986). *Managing the Structured Techniques,* 3rd ed. Prentice-Hall, Englewood Cliffs, N.J.

4

Implementation Planning

Chapter 3 dealt with establishing the context for GIS implementation. The appropriate organizational and planning context was developed, and a vision for GIS implementation created and translated into a multiyear strategy.

This chapter focuses on *implementation planning*. Implementation planning refers to the process of translating the strategy into a series of specific project tasks; when these are completed, the organization will have a functioning GIS. The following categories of tasks need to be accomplished during implementation:

- *Needs analysis.* Assessing the needs of the organization, determining the workings of various business units, their information needs and an assessment of how GIS could be applied to identified work
- *System design.* Determining the processes, data, and technology required to support the needs defined for each identified GIS application
- *Design specifications.* The design requirements determined through the analysis and design process must be translated into technical specifications to be used to structure databases, select software, write custom programs, and select and configure hardware
- *Hardware and software procurement/installation.* The criteria for selection determined from the design specifications are used within a procurement process to acquire the appropriate hardware and software, to be subsequently installed
- *Data conversion.* Data to support the identified applications must be converted from manuscript and other digital formats to the digital format required by the selected software
- *Training.* Staff that will use, operate, and maintain the GIS need to be trained in its use and operation

The discussion in this chapter will provide an overview of these tasks and how to formulate a plan. Later chapters will deal with specific aspects of design and implementation: In Chapter 5 the specifics of conducting a needs analysis is presented, and Chapter 6 deals with the management of the implementation process.

The objective of implementation planning is to produce better systems more cost effectively and sooner. There is a current trend to treat computer systems, particularly small computer systems, as a commodity. Computers may be a commodity, but management and implementation certainly are not. Peter Keen (1991), in his review of information technology use in North America, observes:

> To call key [information technology] a commodity implies that purchasing them is equivalent to using them and that there exist no major blockages to implementation. This encourages the view that it is easy to adopt new technology or to create major new applications. The history of systems development and integration projects clearly shows that it is not at all easy. Indeed, there is a growing sense of crisis in the [information systems] profession about how to dramatically improve the quality, cost, and speed of systems development.

The purpose of implementation planning is to establish the management framework within which implementation will occur to ensure that implementation is efficient, cost effective, and useful to the organization. Implementation planning entails translating the vision determined by the GIS stakeholders into a set of specific tasks that can be implemented within the constraints and guidelines identified during the situational analysis.

The implementation planning process consists of a series of steps. The number of steps and the complexity of each step depends on the scope and complexity of the GIS. However, every GIS implementation requires a plan, and in each plan the following issues must be dealt with.

- *The implementation planning process.* Organizing the tasks associated with implementation and smoothly making the transition from strategic planning to GIS implementation
- *Creating an implementation management framework.* Assigning roles and responsibilities to project participants
- *Developing a conceptual overview.* Articulating the concepts, envisioned during strategic planning, as a set of discrete parts that can be analyzed, specified, and implemented
- *Managing expectations, establishing priorities, and establishing an appropriate sequence of events.* Communicating the requirements for implementation to project participants in enough detail for all to understand what is to be done, how long it will take, what it will cost, and who will need to be involved and for how long
  - *Preparing the implementation plan.* Documenting what has been determined and agreed to

- *Developing "error-free" specifications.* Structuring the planning and design process to produce the fewest number of analysis and design errors
- *Implementing the plan.* Successfully starting out and carrying through to completion

Each of these steps is presented in this chapter.

Upon completion of implementation planning, the following should be in place (not presented in order of importance or implementation sequence):

- A conceptual overview that defines the general business functions that will be supported by the system (these become the GIS applications) and the data that will be converted to support the selected applications. Later, the needs analysis will produce a more detailed assessment of these, but a preliminary assessment is required early for planning purposes
- Consensus on the identified GIS applications—what the GIS will be used for, how, and the expected results
- An implementation schedule by task with appropriate milestones identified
- An accurate assessment and appropriation of resources (funds and people) to complete the project
- Support of the organization for the required resources and allocation of responsibility to complete implementation as defined
- A management and organizational framework for managing implementation of the GIS

The planning process must accurately identify, schedule, and provide the means to manage: people, money, tasks, data, technology, expectations, results, communication, coordination, and policy decisions (requirements for change). The process of preparing the implementation plan forces the resolution of these issues.

The Implementation Process in Overview

The purpose of planning is to arrange the implementation activities into a logical sequence and to schedule time and resources for each activity. It is important to note that parts of the implementation process may be repeated if the GIS is being implemented incrementally. However, the general sequence of activities can be arranged into a series of tasks that approximately follow the process presented in Figure 4.1.

The Implementation Planning Process

The strategic plan described in Chapter 3 will have defined the overall framework for organizing the GIS project tasks and structure. At this point, the

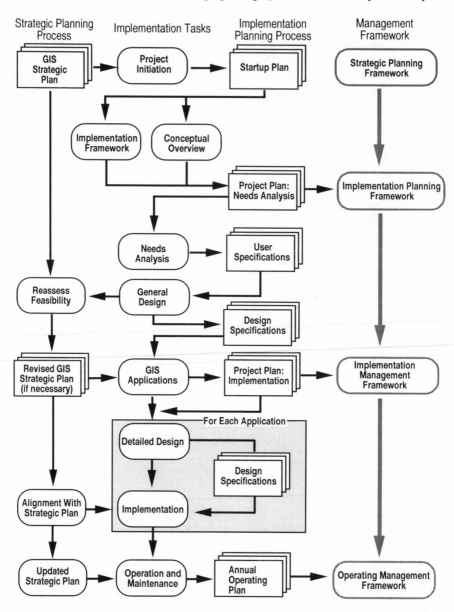

FIGURE 4.1 Implementation planning: stages in GIS design and implementation.

general vision will have been established, major stakeholders and participants defined, and project scope and constraints known. The identified project participants must now be organized into project teams and management committees to undertake the design and implementation tasks.

There is most often a need to review the strategic plan, particularly the

scope and objectives, and roles of various agencies (in the case of multiparticipant projects) prior to preparing the project plan for the next stages of work. A
workshop (or series of workshops for large projects) with project participants
is a good idea at this point. The workshop can provide a review of the strategic
plan, the scope, and the constraints, and present what is to occur next.

The proposed *implementation framework, conceptual overview,* and
implementation planning process as presented in Figure 4.1 should be presented
in a *startup plan.* The startup plan need not be a long document. Table 4.1
presents the contents of a typical startup plan. The startup plan provides written
communication with project participants that can be distributed prior to a
workshop as a basis for discussion. After the workshop, the revised version
becomes a record of what has been agreed to. The purpose of the startup plan
is to bridge the transition between strategic planning, during which everything
is abstract and in the distant future, and implementation planning, during which
concrete issues are dealt with forthrightly. The startup plan can also identify any
outstanding issues that were not resolved during the strategic planning process.
Perhaps most important, the startup plan renews the commitment to a process
for GIS implementation. It has been frequently observed that organizations get
stuck during the transition from strategic planning to implementation, either
continuing to refine the strategic plan or leaping immediately to procurement of
technology. The startup plan provides project participants with an opportunity
to understand what will happen next, why, and how they will fit into the
process, thus ensuring that the whole organization moves forward together.

It could be argued that startup planning is an unnecessary expenditure of
time in an already time-consuming process (in fact someone in your organization
is likely to make this point). However, experience shows that the time spent in
aligning participants with upcoming analysis and design tasks is more than made
up for by the time saved during the needs analysis and the performance of the
design tasks.

Participants will also want to know how they will fit into the implementation
process and what will be expected of them. Middle managers will need to
allocate time for analysis and design activities in addition to the normal
responsibilities of their staff. Remember, the people wanted most to participate
in the process are also likely to be the ones who are the most busy. Informing
them well in advance of the anticipated time requirements will improve your
chances of having them attend analysis and design meetings.

Once implementation has begun, implementation planning proceeds in
parallel with implementation tasks. As work proceeds, more detail becomes
known, permitting more detailed planning. Figure 4.1 shows preparation of new
work plans immediately prior to needs analysis and detailed design of selected
applications. In smaller projects, one implementation plan may suffice. In large,
multiyear projects, more plans may be required. A good rule of thumb is to
perform detailed planning for approximately 3 months, using the startup plan as
context for the overall implementation requirements. The result is a rolling 3-
month plan, identifying specific times for plan review and update. Detailed plans

Table 4.1   Sample Startup Plan Outline

Introduction

- Purpose of the startup plan is to provide an interim work plan and schedule while implementation is being organized
- The plan presents the overall sequence of events and immediate tasks that need to be performed to get implementation started
- E.g., This plan presents what needs to be done to launch implementation, why it needs to be done, who is going to do it, who is in charge of making decisions, what it will cost, and when it will be finished

Background

- A short statement of the strategic vision
- When the project was started and what has been done to date
- Other relevant background information about project participants and working arrangements, particularly in multiagency projects

Scope and Objectives

- Restatement of the scope and objectives of GIS implementation as a reminder
- Restatement of the constraints on implementation - time, budgets, organizational impact, etc.

Participants and Responsibilities

- Who is preparing the plan and initiating the implementation plan? Who is the management authority and what is the interim organizational structure

Task Descriptions

- An overview of all the implementation tasks as per the process presented in Figure 4.1
- Description of the immediate implementation planning tasks
- Milestones and the results to be achieved by each milestone

Schedule

- The schedule for various tasks, start and finish dates, concurrent tasks, task dependencies, etc.
- A Gantt Chart is useful for showing the relationship between task start and finish, and concurrence
- A schedule of the meeting dates, workshops, and other group work is particularly important

Budget

- The overall budget and the budget for specific items such as travel to conferences or demonstrations
- Cost sharing arrangements among participating departments or agencies

---

should be prepared for the implementation task associated with each application after the general design has been determined.

The *general design* is a translation of the user (functional) needs analysis (discussed in Chapter 5) into a description of the system components. The general design is a conceptual model of the GIS from the perspective of information flows and information use. When the conceptual overview was created, some general ideas were developed of how the GIS would be used.

After the general design has been completed, more is known about how the organization functions and what information is processed. It now becomes possible to estimate costs and time requirements better for implementation. It is good practice at this point to reassess feasibility and to ensure that the resources and time allocated for implementation are still realistic. If they are not, either apply more resources or the scope of implementation will need to be reduced. At this point, having a well-supported, active management framework is essential for successful resolution of such issues.

There is considerable variability in what is included in the general design. One approach advocates rigor during general design, creating comprehensive data and process models for the entire organization in advance of designing specific applications. This approach is advocated by most structured system design methodologies. The advantage of the more rigorous approach is that less is left to chance. It is particularly well suited to corporate or enterprise-wide implementations because the database for the entire organization can be designed in advance and then populated incrementally. The main disadvantage is the large up-front cost and the requirement for the benefits of GIS to be taken on faith.

Another approach is to create only a high-level view of information flows and processes and to perform detailed design incrementally. The main advantage to this approach is that substantive activity (buying equipment and converting data) can begin sooner, possibly leading to earlier benefits. The primary disadvantage is the risk that subsequent applications will not easily integrate with later ones. For example, if an early application is to generate data for or share processes with a later application, and the design requirements of the later application are only generally understood, the early application may not create all the required data in the required format, or may operate in a way that excludes sharing processes with the later application.

One way to deal with the uncertainty presented by less rigorous general design is to conduct a pilot project, as discussed later in this chapter and again in Chapter 6.

Independent of the rigor applied to the general design, the process should result in an overall *system architecture*. The system architecture, although still a conceptual view of the GIS, is a technical view defining where data will be located and maintained, which processes will maintain the data, communication requirements (networks), hardware performance requirements, the type and location of equipment to perform required operations, and, in general terms, the software specifications for storage and processing of the data.

Other aspects of the system architecture that may be considered during general design are

- The user interface, the "look and feel" of the system, and how much can be performed by the user without the aid of a technician
- Documentation standards, what will be documented, by whom, and in what format
- Training needs

- Any special environment requirements, such as furnishing, lighting, climate control

The general design becomes the "blueprint" for subsequent detailed design activities. Particularly important is defining process and data needs in terms specific enough to develop the overall system architecture and subsequently the technology selection criteria. These will be used for technology procurement for each application. If general technology specifications are not developed during general design, the technology acquired to implement each application may not be easily integrated into a complete system.

Having confirmed the general design and the selected applications, a logical sequence for implementation can be developed. There are many elements to be considered when determining the sequence for implementation of identified applications.

Applications that will *create data* needed by other applications (e.g., the creation of the map base for subsequent themes of information) should be implemented first. It is generally better to begin with a *simple application* while the staff are still learning and bugs are being worked out of the system. Also, more complex applications can often be built by assembling several simpler applications, reducing later risk and cost. *A suitable return on investment* and/or importance to the organization should also be considered, thus ensuring that selected applications will adequately demonstrate the benefits that have been promised for what is now likely many months. An early return on investment helps to ensure continued support for later applications. A more detailed discussion of application implementation priorities is given in Chapter 5.

Each selected application must then be designed in detail. *Detailed design* consists of translating the functional specifications and system architecture, describing what the system will do into specific technical descriptions of how the system will perform the required functions for each selected application. The amount of work comprising the detailed design largely depends on the approach used for the general design, as discussed. If a very rigorous general design study was performed, then the detailed design will mainly consist of extracting and refining the specifications relevant to the specific application. If the general design was indeed general, then detailed design will require detailed analysis of data and processes. Either way, the detailed design must provide a database design (how the data will be stored, accessed, and maintained), a process design (how the data will be manipulated and analyzed), and specification of how the application will fit into the workflow of the user (how the application will be used to perform day-to-day work). The process of detailed design and some of the associated techniques are presented in Chapter 5.

Finally, the completed design is converted into a working system. While the discussion to this point has stressed the importance of comprehensive planning and design, it is equally important to remember that a design, however comprehensive, is not a working system. Managing implementation is itself a significant topic, to which Chapter 6 has been devoted.

Having reviewed the process in overview, let us focus on some specific aspects of implementation planning in more detail.

Creating an Implementation Management Framework: Assigning Roles and Responsibilities

Management framework is a term used to describe the organizational structure within which implementation will occur. It consists of a hierarchy of communication and decision-making forums (usually some form of committee), which will ensure implementation is participatory yet efficient, cost effective, and useful to the organization. In defining the management framework there are three main factors to be considered: what decisions need to be made; who can best make them; and how to organize the participants to best facilitate the activities associated with those decisions.

The decisions to be made can be broadly generalized into the following categories:

- Policy, budget, and strategic direction decisions
- Implementation management
- Technology selection
- Technical implementation alternatives

The participants likewise can be categorized as follows:

- *Sponsors,* which can include senior management of the participating agency(s), and outside agencies that may be providing additional funding (e.g., another level of government or a venture fund)
- *End users/clients* of the new system, the people currently performing the business functions that will be served by the GIS
- *Project management,* who have been the responsibility for managing planning and implementation
- *System designers,* who determine and translate end-user and organization needs into technical specifications
- *System implementers,* from participating agencies identified to participate in the implementation and outside specialists retained to perform specific implementation tasks
- *Data providers,* which may include the end users of the GIS, existing computerized systems, outside agencies, data companies, and contracted services

These persons and agencies will not all have the same stake in implementation and they will not all be involved in all stages of planning and implementation. Therefore, different organizational structures and mechanisms will best suit different participants.

Structure options include:

- Semipermanent (i.e., for the duration of the project) management (steering) committees that serve as the management authority for the project
- Special-purpose management committees for resolving special policy- or change-related issues
- Liaison committees, which provide for communication with peripherally involved participants (e.g., data providers, interested agencies)
- End-user advisory committees, often the working-level equivalent of the management committee, that provide access to staff and information about business functions, and that resolve implementation issues
- Special working groups, charged with solving a specific problem or undertaking a specific task series within the overall implementation plan (e.g., conversion of a selected data set)
- Workshops, which may include participants from any of the other structure mechanisms, usually about a day in length, and often as part of the analysis and design process

How these are applied and combined becomes a function of the needs of the GIS implementation, including

- Whether the project is a single or multiagency project and the relationship between participants
- The size and complexity of the GIS implementation
- Organizational culture
- The amount of work to be performed within the agencies versus that contracted out

Successful decision-making during implementation planning is achieved through the appropriate combination of structures, participants, and roles. A hierarchy of structures and roles will help to ensure that the best decisions are made at all planning levels. Table 4.2 presents an example of a management framework for GIS implementation. Note that it includes definition of the responsibilities, the reporting structure, meeting frequency, and participants. This latter point is very important. If participating agencies are permitted to move representatives around in the framework through delegation, for example, then the framework breaks down. Specifically, if one agency executive manager decides not to attend management committee meetings and delegates his position on the committee to the working group member, then soon other executive managers will do the same. The end result is that the management committee and the working group consist of the same members, destroying the hierarchy of responsibility. The impact of this breakdown in structure is not usually felt immediately, but the effects will become very evident when a major issue needs to be resolved. The solution to this problem is to keep executive involvement to

Table 4.2   Sample Management Framework

Framework Element	Participants	Responsibility	Meeting Structure and Frequency	Reporting
Management authority (steering committee) sponsors	Policy makers (department heads) of sponsoring agencies	Policy decisions Approval of plans and resource allocation Conflict resolution	Quarterly More frequently at key points or or as issues dictate	Executive or individual agencies or elected officials
Liaison committee	Management representatives from interested but not sponsoring agencies	Project monitoring Communication Coordination	Quarterly, or as issues resolution require	Executive or elected officials of each agency
User working group	Line managers and senior professionals of sponsoring agencies	Facilitate needs analysis and design studies Facilitate/review project plans and specifications	Weekly to monthly, depending on work load and project phase	As a group to the project management team Individually to their management
Design and implementation	Project leader or manager Internal and contracted analysts and programmers	Tech design specifications Application construction System testing	Daily to weekly Frequent short meetings	Project management team
Project management team	Project manager Project leader(s) End-user appointments	Project plans and deliverables Coordination GIS implementation	Daily to weekly Frequent short meetings	Management authority

acceptable levels, to disallow delegation (or at least not to working group members), and to ensure that management committee agenda topics are appropriate to senior management.

The Role of Senior Management

The GIS community has spent considerable time analyzing and discussing the essential elements for successful GIS implementation. One of the elements that repeatedly stands out is senior management commitment. The appropriate role for senior management can be debated, and how extensively management is involved in a particular project is probably more a function of organizational culture than of any other factor. That said, as a minimum senior management should be seen to be interested in the GIS and should ensure appropriate commitment in the level of management assigned responsibility for GIS implementation. To implement the GIS effectively requires management discipline, time, resources, and skill. These need to be created, fostered, and facilitated by senior management. Senior management will likely also be the project sponsors and as such will be required to approve expenditures and any changes to organizational structure or policy. This means that senior management will require at least a general knowledge of the GIS implementation process and the impact of GIS implementation. The strategic planning process should have created the environment within which these will naturally evolve during implementation. If they have not or if they need re-enforcing, the appropriate environment can be created during the establishment of the implementation management framework.

The GIS Champion and Implementation Management

There was a time in GIS development when it was widely accepted that a GIS project had to have a "champion." A champion is someone entirely committed to the idea of implementing a GIS within their organization. Champions tirelessly pursue the goal of GIS implementation by selling the idea to management, co-workers, and anyone else willing to listen. It was widely accepted that without a champion an organization could not successfully implement a GIS. This view is changing as GIS becomes more widely known and accepted, and as GIS planning becomes part of information technology planning within organizations. The need to introduce and sell GIS as a new technology is disappearing and with it the absolute need for a champion. This is not to say that organizations should ignore or discourage individuals with an overwhelming desire to implement GIS; such energy and commitment is of obvious value. However, increasingly organizations are implementing GIS within multiparticipant contexts. Consequently, the GIS champion is evolving into a project manager and coach. This is a sensible view. If past thinking was that a champion can accomplish a lot, then a committed team should be able to accomplish more.

This raises the question of the appropriate role for the manager or team leader during implementation. There are several possible roles:

- *Politician/salesman.* Promoter and proponent of GIS, selling the vision and securing the necessary resources for GIS implementation
- *GIS chief administrative officer.* As manager of the resources and tasks associated with GIS implementation
- *Technical expert.* The chief architect of GIS design, perhaps designing the database and creating the applications
- *Coach.* Team builder and leader

Each of the preceding roles possesses strengths and weaknesses. The ideal may in fact be an appropriate combination of all of them, depending on the needs of the organization. However, such individuals are rare, and few organizations will be lucky enough to have someone with all the desired traits leading their GIS implementation. Often the person emerges rather than being selected, depending on how GIS implementation began.

In some instances, the GIS champion emerges as the politician/salesman. In this role the champion is very successful at creating enthusiasm and initial support, but may lack the skills required to bring the GIS through to implementation. Several examples exist of organizations in which tremendous enthusiasm for GIS was created at the outset, but few results were realized.

After the vision has been created and plans have been formulated, one of the roles of the implementation manager by necessity becomes that of administrator. In larger GIS implementations, an administrator may be assigned, but in most instances the manager will need to perform associated administrative duties. However, if too strong an emphasis is placed on the administration role, the GIS implementation may suffer from lack of leadership and technical direction.

In other organizations, implementation begins as a technical task. An inexpensive GIS system is purchased, and a technically oriented individual is assigned responsibility for creating results—the manager as technical expert. This may occur within the context of a pilot project, or this approach may reflect the needs and plans of the organization at the time. If the GIS expands in scope and in number of participants with a technical expert as manager, the manager may attempt to impart a singular view of GIS on participants and applications. That view may be counterproductive to implementing a GIS that satisfies the needs of all participants. The danger is that the technical expert will not be able to make the shift from providing solutions to identifying needs and evaluating alternatives.

When GIS implementation involves more than one department or agency, and the GIS will support several activities, one of the most important roles for the manager of GIS implementation is that of team builder and leader. Large implementation projects involve many people with diverse backgrounds. A successful project manager must be as comfortable dealing with senior management and end users as with technical systems staff.

The role of the manager also changes as GIS implementation moves from planning to implementation management to operation, requiring different skills at various stages. Therefore the subject of the project manager is dealt with again in Chapters 6 and 7.

Building Project Teams

> Committees have agendas; a team has a vision.
>
> RALPH SLAYER

Increasingly, successful management means building successful teams. A measure of a successful team is whether participants are compliant or committed. Compliance means that participants will show up as directed, provide inputs as instructed, but otherwise will volunteer nothing. At best, the success or failure of the GIS means nothing to participants. At its worst, resentment towards the GIS and perhaps towards the manager means that participants look for ways to sabotage the project. Commitment means that participants enthusiastically participate in GIS activities, readily offering new ideas and actively seeking ways to achieve success. Compliance results from edict; commitment from shared vision and a sense of belonging. A well-functioning team does not need to be supervised or motivated; it meets virtually any challenge because it wants to.[1]

To a large extent the difference between a compliant and a committed team rests with the team environment and its support from the organization. As discussed in the previous section, the manager, as team leader, has a large effect on team performance, and therefore much of the teams' success or failure depends on selecting the right manager at the outset. However, there are many ingredients to building successful teams.

Assemble the Right People for the Right Job. Either match the requirements of the implementation plan with the skills of staff to be assigned, or ensure that required skills are taught to participants. It is frustrating and threatening to be assigned responsibility for which one does not have the required skills. Frustration soon gives way to resentment and compliance. Participants should be organized into separate teams with similar skills and responsibilities, unless there is a specific reason for mixing them, and in that case they should have a shared purpose for working together. Forcing technical design staff and end users to participate in joint discussions that one or the other poorly understands and that are of little interest to the other, will again result in frustration. It is not necessary for each participant to understand all aspects of implementation, only that the implementation staff as whole understand.

Communicate the Needs of the GIS Activity up Front. As with the first point, people are most comfortable and productive when they understand what is expected of them. This applies to both the *type of work* and the *results,* and to the *time commitment required.* Staff assigned to the GIS project will also have other work. It is important to *balance GIS duties and other job responsibilities* properly so that staff do not feel that the GIS is being implemented at their expense. This may require the temporary assignment of some staff, and negotiation with staff managers for reduction of other work for others. This means that both *personal and organizational commitment* is required.

Create an Environment of Trust. This is accomplished by openly sharing information throughout the implementation process. Start up new activities and working groups with a motivational talk from senior management. Present the vision for GIS in the organization as background for specific objectives and tasks. Do not create false deadlines or hide important details that could later be revealed as deceptions.

Once working groups have been established, it has proven valuable to provide bonding opportunities through group activity. This can include planning retreats and extracurricular activity, but the simplest and most effective is eating together. Popular wisdom, well supported by anyone who has led a team, says that you cannot have an effective team if they do not eat together.[2] There is something about the basic instinctive process of "sharing bread" that brings people closer together. This can be as simple as providing a lunch space or more formal special get-togethers. A meal to kick off a new team is very beneficial.

Teams must of course also have purpose to be of use to the organization and the GIS implementation, but this becomes more a process of facilitating and leading than commanding with direct authority. "The purpose of a team is not goal attainment, but goal alignment" (DeMarco and Lister, 1987, p. 126). When an effective team has a shared purpose, it is directed and driven to the result. Goal attainment is a given.

Sometimes it is useful to create a temporary common working environment, often called a "skunkworks."[3] The decision to create a special working environment depends on the working environment, the organizational culture, how much the new GIS processes will differ from current work processes, and how many of the team are seconded or contracted. There are differences of opinion about how useful these are. The benefits are creating a strong focus and separate team culture. The disadvantages can be the creation of elitism and resentment from staff not involved in GIS implementation. The decision to use a skunkworks approach should carefully consider the benefits and consequences within specific organizational contexts.

Existing teams lacking motivation or direction can frequently be stimulated with the introduction of outsiders such as consultants or staff from another part of the organization. However, the situation and purpose must be clearly understood by all so that any sense of manipulation either of the team or the addition is avoided. Outsiders may of course also be introduced to bring missing skills or to help overcome periods of extensive work. When possible, such additions should be identified in advance and supported by the existing team members to prevent upsetting existing working relationships.

Although most of the time technical staff and end users will work in separate working groups, at various stages in the GIS implementation it is worthwhile to bring two or more teams together for joint work sessions. Technical design teams and end-user working groups should meet at strategic points to develop or refine specifications prior to implementation, to review prototypes, and again as individual pieces of the system are completed. These sessions are becoming commonly known as joint application design (JAD) sessions.[4] During these sessions functional requirements, technical problems, and potential solutions are discussed as a group. When applications are being designed to satisfy the needs of two or more organizational units, end users frequently disagree on the definition of the problem and proposed solutions. Many such problems can be resolved immediately by bringing technical staff and end users together, saving enormous amounts of time for both groups. These sessions are often controversial and heated, and an experienced facilitator is often required to ensure positive results.

Thus far the discussion has focused on the dynamics of assembling successful teams; however, the needs of the project and the organization must also be considered in assembling teams. Issues that must be addressed include the following:

- Ensuring that all organizational units are adequately represented
- Assembling an appropriate mix of skills and experience to complete the required tasks
- Ensuring that the organization is receiving good value for its expenditures on human resources through the selection and assignment of staff and contractors
- Facilitating staff, who will later be required to operate or manage parts of the system, to be involved in design and implementation to ensure adequate knowledge and commitment

The number of teams and the size and composition of each team is determined by the scope of the project, the size of the organization, and the level of involvement required from its various parts. Also, different parts of the organization will be affected at different times. In an incremental implementation, once the general design activities have been completed, only end users in business units that support the application currently being designed and implemented will be affected. This means that with careful planning, staff can often be rotated to cover for staff participating in application implementation. Determining the sequence of development and related staff assignment requires an overview of the entire GIS from which specific applications can be selected and scheduled for development.

Developing a Conceptual Overview

With the implementation management framework established, the implementation tasks can begin. The first analysis and design task and the first challenge for the working groups is to convert the vision, objectives, and constraints, as defined in the strategic plan, into a high-level model of the GIS. This is not as complex as it may appear at first glance. The purpose of this exercise is primarily to create a shared understanding of the general shape and size of the GIS once it is operational. Remember the conceptual overview will be followed by several increasingly detailed analysis and design tasks that will refine the conceptual model developed at this point. Another way to look at this task is as an educational activity. Through presentations and discussion, the participants as a whole and at all levels develop a reasonable understanding of what is being built and how it will be done. This activity could be described merely as a set of seminars, but experience has shown that the activity will be much more successful if participants have a purpose. This is also an opportunity to further reinforce the sense of ownership and commitment to the GIS through participating in its conception.

Many techniques can be used to facilitate the conceptual overview process. For example,

- Have the team attend a related conference to learn what others are doing and to review exhibits from GIS vendors
- Conduct in-house seminars using a GIS consultant or other expert to present some alternatives and lead subsequent discussion
- Conduct site visits to organizations performing similar work that are now using a GIS
- Have participating departments or agencies present overviews of their current business functions and initiate discussion of how a GIS might be used to support the business functions
- Conduct a demonstration project using data from some of the organization's business functions as an example of what could be done

The preceding can be used in combination with each other, and there are many other techniques that can be employed. Note that in using any of these or similar techniques to stimulate ideas, the group should be cautioned against wanting to copy any specific example or to leap to conclusions about technology. For that reason it may be worthwhile to have a meeting prior to starting any of these techniques to explain the purpose. Some individual or a group should also be charged with documenting the findings and discussion, as a basis for creating the conceptual overview. At some stage an individual will need to prepare a document for subsequent discussion by the group. Groups do not easily prepare documents.

Organizations will often look for expert advice and support at this stage. In the early stages of implementation most organizations do not yet have a body of expertise from which to draw. Additionally, the work associated with formulating a conceptual overview can be daunting for most staff. There is usually a sense of impatience at this stage as well. The organization wants to get on with it, and the use of a consultant can accelerate the process.

There are four elements that must be dealt with generally in the conceptual overview, but in an interrelated manner:

- The existing physical organizational structure
- The business functions (mandates) of the organization and the subsequent business decisions made by the organization
- The information that the organization uses to support the business functions
- The technology that could be employed to handle the data and information

In relative terms, organizational structure and technology tend to change more often than business functions and the information required by the organization. For this reason, and others that will be dealt with later, the process should focus on the primary activities of the organization and on the information needed to support related decision-making. The technology to support the organization's function will be determined later, as will any necessary changes to the organizational structure. At this stage it is not necessary to determine how the information is created, or what the processes are or will be required to process it; the intent is to identify which business functions will be included in GIS implementation and, in general terms, what type of information will handled by the GIS.

Some facts about the organization should be available from the strategic plan. The existing organizational structure, the units comprising it, and descriptions of governing legislation, mandates, critical success factors,[5] and equivalents contained in the strategic plan can be used as a starting point for defining business functions. Each of these can subsequently be defined in terms of the major decisions made and the information required to make them. Each organizational unit that will use the GIS will require access to computer resources. Organizational units that create and maintain information may require more complex workstations (e.g., digitizing tables) than units that are solely information users. The number of people who need computer access in each organizational unit and the type of work they perform roughly defines the number and type of workstations. The distribution of workstations in turn defines the communication requirements in general terms. Assembling this information about the organization, its business functions, the information needs, and an approximation of the computing technology produces the required conceptual overview of the organization.

Creating a conceptual overview for the GIS is made more complex by the current activities of the organization. Most organizations will have related projects and technology that must be considered when developing the conceptual overview of the GIS. Typical issues that are encountered include the following:

1. The new technology is being implemented to support old methods.
2. The new technology, and possibly new methods, may need to address the concerns of many disciplines and departments whose requirements conflict.
3. Most organizations have existing systems and technology that must somehow be used, integrated, or migrated from.
4. There are often projects and systems under development that must be worked into the planning.
5. The need to satisfy immediate objectives and provide immediate payback must be balanced with long-term stability and strategic objectives.
6. Corporate and departmental objectives, both of which must be satisfied, may conflict.

At this stage in implementation planning, many of these can only be noted as issues to be dealt with as implementation proceeds. It is important not to become stymied by these issues. However, it is equally important not to ignore them. Deal with any that can be dealt with, and include the others explicitly in the implementation plan.

Selecting Applications

The conceptual overview now provides a framework for identifying system applications. *Applications* are the end result of applying information technology to a business function. They provide computer access to information needed to support the business function. They may also computerize some aspects of the business process in addition to the data, providing a higher level of support than just information storage and retrieval. An example is computerizing the allocation of forest stands for wildlife or recreation use, rather than just computerizing the mapping of forest inventory.

Considerations in selecting applications include: critical success factors and mandates, technical risk, availability of data (conversion requirement/cost), position of the application in a logical implementation sequence, level of user support, and the return to the organization (either tangible or strategic), as discussed earlier. These considerations can be used as criteria for determining which applications should be undertaken first. Significance can be applied to each criterion, creating a weighting system, if some criteria are believed to be of greater importance than others.

Figure 4.2 presents a diagram showing how these criteria might be used to rank applications for selection. A success quotient or value can be assigned to each application by applying the criteria to candidate applications. In the first example, words defining a three-class subjective rating are used to evaluate each candidate application. In the second example, a numerical rating system is used, and weights are assigned to each criterion reflecting their relative importance in selecting the applications. Other variations on these themes are possible. This process is particularly useful when several agencies or departments each have proposed candidate applications. If the criteria (and significance weights, if applicable) are determined in advance by all participants and then applied as a group, the process provides equal input by all participants.

At all stages it is important to differentiate winners from losers in terms of the value of the application to the organization. However, it is even more important to do so in the early stages of implementation, because doubts about the benefits from GIS will still exist. Applications selected for well-conceived reasons must be differentiated from those based on wishful thinking and ill-conceived ideas. Technology must enable something that people really want and need, not something they should logically want and need. The technology should remove barriers to personal and organizational productivity. It should build on the day-to-day activities of existing business functions while opening the door to new activities that are the natural evolution of current functions. It should not

CRITERIA	PARCEL MAPPING	BUILDING PERMITS	ENGINEERING DESIGN
CSF/ Mandate	Low	Moderate	High
Return to Organization	Low	High	High
Data Available	Yes	Yes - extensive conversion	Yes - extensive conversion
Technical Risk	Low	Moderate	High
Prerequisite for Other Applications	Yes	No	No
Level of User Support	High	Low	High

CRITERIA	SIGNIFICANCE WEIGHTING	PARCEL MAPPING	BUILDING PERMITS	ENGINEERING DESIGN
CSF/Mandate	2	(2) X 1 = 2	(2) X 2 = 4	(2) X 3 = 6
Return to Organization	2	(2) X 1 = 2	(2) X 3 = 6	(2) X 3 = 6
Data Available	3	(3) X 3 = 9	(3) X 1 = 3	(3) X 1 = 3
Technical Risk	3	(3) X 3 = 9	(3) X 1 = 3	(3) X 1 = 3
Prerequisite for Other Applications	3	(3) X 3 = 9	(3) X 1 = 3	(3) X 1 = 3
Level of User Support	1	(1) X 3 = 3	(1) X 1 = 1	(1) X 3 = 3
TOTALS		34	20	24

FIGURE 4.2 Criteria to rank candidate applications.

produce dramatically changed information products or change the purpose of a business unit unless that was the explicit purpose. Failing to select appropriate applications can end GIS implementation in its early stages.

A significant challenge in selecting early applications is trading off applications that have high early payback against those that must be developed to create the infrastructure for implementation of later high-payback applications. Unfortunately, there is no simple solution to this dilemma; it can only be resolved through discussion, consensus, and, finally, a decision by the management authority. The topic of selecting applications is more fully discussed in Chapter 5 in the context of needs analysis and design.

Managing Expectations, Establishing Priorities, and Establishing an Appropriate Sequence of Events

Maintaining support for a GIS implementation requires a delicate balance between enthusiasm and reality. To create interest it is necessary to portray an enticing vision of the future that focuses on benefits and exciting new capabilities. However, to ensure continuing support for GIS implementation, this view must be balanced with a realistic description of costs and time requirements. This is perhaps the most difficult aspect of planning and managing the implementation process. The qualifiers that were attached to preliminary cost estimates are quickly forgotten. The various agreements by user departments to postpone development of their applications until other work has been completed can fade unless there are constant reminders throughout the implementation process. Management of expectations and results must begin at the same time as implementation begins, and must be continued until after the system is operational.

If the process to this point has involved the key stakeholders (management, end users, outside sponsors), then the issue of expectations will be more easily dealt with than if it has not. Startup meetings should include a review of the objectives and constraints attached to the project. Agreements on the selection of applications, data to be converted, technology to be purchased, staff training, and especially cost sharing must be documented and signed off by the management authority. These can all be included in the implementation plan. In fact, documenting the agreed to sequence for implementation becomes one of the main purposes of the implementation plan.

Another aspect relates to determining realistic time expectations. In general terms the time required for GIS implementation (T) is:

$$T = \text{time to build the platform}$$
$$+ \text{time to build the applications}$$
$$+ \text{time to develop and learn new procedures}$$

This formula does not include the time for up-front implementation planning or time spent as a result of administrative constraints, pilot projects, or other interim steps. A reasonably accurate estimate of the implementation time can, however, be determined by summing estimates of each element in the equation.

In managing the expectations of end users and management, it is important to explain adequately what takes time and why it is worth waiting. Many projects are perceived to be failures because expectations were not properly established at the outset. Lacking documented and understood expectations, projects are perceived as failures because they appear to have delivered less, or taken longer, or cost more than was expected.

A particularly difficult situation may arise when the original proponent of GIS has oversold the project and created unrealistic expectations in the minds of end users and management. This is particularly problematic when the original proponent is replaced by another manager. In these circumstances the new manager has few options. Expectations must be re-established to reflect reality. This usually means retracing some steps and re-establishing credibility with stakeholders. It is likely that the problems were in part created due to the absence of a comprehensive implementation plan. Credibility and realistic expectations can be established by providing a substantiated implementation plan.

Preparing the Implementation Plan

Documenting the plan can seem to be a large and onerous task and therefore is often not done or is done quickly. However, a well-documented implementation plan is perhaps the best investment of time in the entire GIS implementation process for the following reasons:

1. Developing a detailed implementation plan forces the consideration of many issues before they arise, thereby ensuring consensus on action and the availability of necessary resources.
2. The plan creates a commitment to an agreed-upon process and schedule of work—especially important in multiparticipant projects.
3. A well-developed plan and task schedule ensures that all project participants are aware of their responsibilities.
4. It builds the confidence of project sponsors (senior management).
5. It can provide early warning of potential problems and provide the basis for understanding the implications and identifying solutions.
6. It permits the project manager and related staff to proceed confidently, improving productivity and project moral.
7. Activities can proceed concurrently, when appropriate, permitting a faster pace for development without increasing risk.

The task of preparing the implementation plan can be simplified by using a template for the plan. An example of the table of contents for an implementation plan is shown in Table 4.3. Specific activities (workshops, meetings) can be held to deal with each aspect of the implementation plan. The meeting notes and records of agreements reached provide the substance for the plan. Other information, such as the scope and objectives, background on the history of the project, past involvement of persons and agencies, can provide continuity and serve as reminders of past decisions. These will be available from earlier documents, if documentation has paralleled implementation planning.

The plan can only provide details for tasks that are well understood. Tasks that are not well understood or are farther on the planning horizon can be dealt with by scheduling a separate planning task to produce the required details. In this manner, more of the plan can be scheduled without guessing at important details. As more of the project and system details become known, and staff

Table 4.3   Sample Outline for an Implementation Plan

Introduction

- The contents and intended audience of the report
- The purpose of the startup plan in relation to other project documents (that is, it follows the strategic plan and precedes a detailed implementation plan)

Background

- A short statement of the strategic vision
- When the project was started and what has been done to date
- Other relevant background information about project participants and working arrangements, particularly in multiagency projects

Scope and Objectives

- Restatement of the scope and objectives of GIS implementation as a reminder
- Restatement of the constraints on implementation: time, budgets, organizational impact, etc.

Conceptual Overview

- An introduction to the conceptual overview describing the business functions that will be supported and how; the data that will need to be converted; the organizational units that will be affected as a result; and the overall sequence that development will follow (i.e., who will be affected first and why)
- The use of diagrams to relay complex relationships is very effective

The Management Framework

- Define the participants in terms of both organizations and levels within the organization
- Define the committee and working group structure proposed or previously agreed to. The management committee established during strategic planning may continue or some new structure may be created
- When known, indicate committee members and other project participants by name
- Define the responsibilities and authority of each committee and working group. When known, identify the chair of each group

Task Descriptions

- Provide a general overview of the process that will be followed during implementation, such as that presented in Figure 4.1
- Describe each of the major steps and activities in the process. The use of phases or task series can be useful for explaining complex tasks without needing to detail individual component tasks

Schedule

- Discussion of the overall time frame for implementation: could be months or years. At this state the broad time frames should be dealt with. Large projects may be broken into phases, providing general time frames (such as fiscal or calendar years) for each phase, indicating the purpose and results of each phase
- A Gantt chart showing the project phases or task series is very helpful and easily understood by most people

Budget

- Present the current, committed, and planned budgets for the project. This portion may only be communicated to management depending on the practices of the organization(s). Generally, if permitted, it is useful to communicate the project budget to all participants as an additional parameter governing the rate of progress

Administration

- Description of how the project will be administered including: management authority (what can be decided by the project manager, and what must go to management committee); administration of funds (particularly in multiagency projects); personnel management (particularly arrangements for secondment of staff); contracting practices or restrictions; information technology architectures, plans or standards that must be adhered to; and other administrative policies, restrictions or special dispensations. The policies and practices need not be stated in the startup plan, just referred to as governing project activities become more familiar with the project, less time will be required for task planning.

Project management software provides a useful tool for developing, maintaining, and monitoring plans. Project management software is available for almost every type of computer system and varies in price from a few hundred dollars to several thousand dollars. Project management packages available for microcomputers will provide most projects with everything needed. The features of these systems varies somewhat by package, but generally include various reporting capabilities, personnel and financial management planning, in addition to schedules and compatibility with various other software. At the root of all these systems are network diagrams or the project evaluation and review technique (PERT) charts. These provide the implementation planner with the ability to define and schedule tasks, to place tasks in sequential order graphically, to allocate resources, and to determine which tasks must be completed before others can begin (dependencies).

Figure 4.3 shows a sample of a network diagram. Determining task dependencies is very useful in planning a project. The necessity to define dependencies forces the project planner to consider relationships between various tasks—which tasks can be performed concurrently, and which tasks are dependent on completion of other tasks. The process of preparing the plan is often as useful as the plan itself because of the need to consider in detail the many relationships. Most project management systems include the ability to allocate resources (people, equipment, money) by task and to determine when each resource has been fully utilized. Following this process thus allows not only planning and scheduling, but checks on budgets and staff allocation as well.

The example in Figure 4.3 is very simple and could of course easily be planned without the use of a network diagram. However, the technique gains in value when hundreds of tasks in multiple task streams are to be planned. This process was originally created for development of the Polaris missile and has been used on countless complex projects since.[6] Also, only the task number, task name, start and finish dates, and the person responsible for completing the task are shown in this example. Many project management systems facilitate the display of earliest and latest dates, slack between tasks, task descriptions, and various codes. The incorporation and display of codes can be a useful feature, as these can be related to budget codes in a separate spreadsheet to facilitate project accounting.

Some other useful rules of thumb can be followed when preparing project plans:

1. Include *frequent review points,* especially at the beginning of the project to demonstrate that the project is on schedule and to build confidence among project participants. Most of these can be conducted with the project team, but some should also be conducted with management.
2. Make *no individual task* (box in the network diagram) *longer than 5 days.* If a task cannot be performed in 5 days, then the task should be

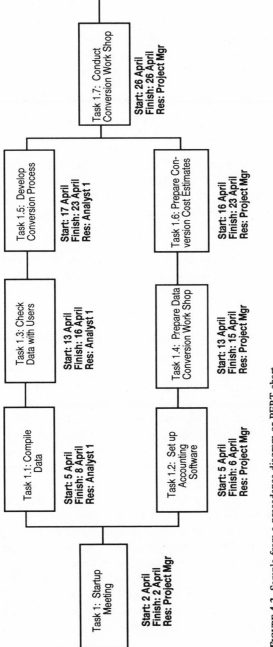

FIGURE 4.3 Sample from a precedence diagram or PERT chart.

broken down into component tasks. If the task cannot be subdivided into 5-day segments, then the task is not well enough understood and should be studied further prior to being performed. Another reason for the 5-day rule of thumb relates to staying on schedule. If a 5-day task slips, the most that can be lost is 5 days. The impact can be determined immediately and remedial action taken. Five days can be made up relatively easily. If the task is 30 days long, slippage will not be determined until the deadline for the task approaches. Thirty days is difficult to make up in 3- to 6-month projects.

3. Ensure that *for each task there is a person responsible.* Other resources may be assigned as well, but for each task one person should be identifiable as being responsible for its completion. If that person is to be recruited or contracted, ensure that there are preceding tasks dealing with appropriate administrative tasks, and that adequate time has been allowed.

There are many other aspects to implementation management; these are discussed in Chapter 6. For now, let us proceed with items to be considered during implementation planning.

Developing "Error-Free" Specifications

A question that is frequently asked is: How can I shortcut the system design process? I understand the benefits of planning, but my organization is impatient, and we need to get on with physical implementation now. The answer is, one can spend the time now creating a well-planned project, or spend the time later redoing what was not well understood or poorly designed. It is interesting to observe how many organizations do not have time to do it right, but they have the time to do it over. There is no shortcut to quality. Clearly an "engineering quality" approach will take longer, but the result is a more usable, more reliable system that will not need to be rebuilt or replaced as technology changes.

There is an unfortunate modern myth that there is no such thing as "bug-free code" or defect-free systems.[7] This does not need to be. However, quality (as a measure of being free of defects) cannot be added after the system is built. In fact, it cannot be added after the specifications have been written, and ideally, quality assurance should begin during planning and team building. The objective should be to create a "cult of quality" wherein each team member and each task emphasizes quality. If quality is defined as *conformance to requirements,* the concern for quality at all stages in design and implementation becomes apparent.

"So what's the big deal?" one may ask. "A few errors can be fixed as we proceed, and some changes will be required no matter what we do." Well, recent studies of hundreds of implementations demonstrate that it *is* a big deal and that large savings are possible.

Figure 4.4 shows a simple representation of the relationship between the

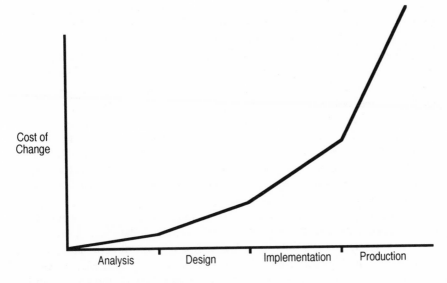

FIGURE 4.4 The cost of fixing a defect.

cost of fixing a defect and the point during implementation at which it is fixed. This simple diagram is supported by the finding that each dollar spent on system development generates on average an additional 60 cents of expenditure per year for operations and maintenance (Keen, 1991, p. 141). In other words, a $1 million expenditure today can be expected to generate an additional $3 million of expenditure over the next 5 years (the current life span of most systems). In light of these facts, reducing subsequent expenditures for operations and maintenance takes on new significance.

The quest for quality begins with the message from management, and where the emphasis is placed—on quick delivery and meeting schedules at any cost, or on the pursuit of quality. Once the direction is established, planning, specifications, and implementation must follow through for quality objectives to be achieved.

An issue that frequently influences quality and the ability to deal confidently with design and implementation issues is uncertainty. Uncertainty results from dealing with the unknown or the poorly understood.

Dealing with the Unknown or the Poorly Understood

As implementation proceeds, and as planning identifies a sequence of tasks to be performed, individuals are assigned to perform the tasks, technology choices are evaluated, concepts of what the system will do and how it will work are formulated, and plans for integrating the new GIS begin to solidify. At some stage in this process, it will be realized that there are many aspects of GIS implementation that are not well understood. At this point, specific measures must be taken to ensure that inadequate management of uncertainty and risk do not jeopardize the GIS implementation.

In Chapter 3 we identified three significant variables that must be dealt with in successfully matching organizations to technology: complexity, uncertainty, and interdependence.[8] We learned that the greater the diversity, uncertainty, and interdependence associated with performing a given type of work, the more complex the process for successful introduction of the new technology and the greater the need for management of structure. In Chapter 3, our interest was in the impact of these issues on organization strategy and creating environments receptive to new technology. These issues arise again in this discussion of implementation, but this time in the context of specific actions to manage uncertainty and risk during implementation.

Advanced forms of information technology such as organization-wide applications of GIS have broader and deeper organizational consequences than single-agency, single-application GIS. Increasing the functionality of an information technology increases the levels of learning and adjustment required to utilize it, ranging from operator skills to changes in organizational procedures and structures, to cultural fabrics. More sophisticated applications change not only the mechanisms for information handling, but organizational structure and job functions as well. All of these will create uncertainty, risk, and, in the staff responsible for implementation, stress.

Some aspects of these issues are readily manageable, while others are not. Examples of sources of uncertainty that can be successfully managed are

- Lack of specific knowledge or skills to perform given tasks—these can be acquired
- Response of staff to situations of uncertainty (stress)—the work environment can be structured to reduce the stress for individual workers and provide reassurance
- Technical problems related to data conversion, technology integration, functional capability, and related others—performance measures can be established and tested

Examples of others that cannot be managed or that are more difficult to manage are

- Technology change—trends can be monitored, but it is difficult to attain any degree of certainty with technology choices
- Changes to the driving forces of an organization and hence business direction and priorities for information systems application
- The psychology/sociology of users—the effect of new technology on working relationships between individuals and departments

If uncertainty is not effectively managed, it will surface as poor project staff morale, a series of real or apparent technical failures, unhappy end users, and poor overall returns on GIS investment. At this point in implementation, the management authority should request that the sources of uncertainty and risk be explicitly identified, along with the plans for dealing with it. If the management authority is not satisfied that apparent sources of uncertainty have been identified, or that the proposed action of dealing with it is not adequate, then the management authority should take separate action to deal with the issues (e.g., retain an expert, involve other management staff).

Technical risk has an added impact on cost and cost estimating. High levels of uncertainty about the success of an application may lead to over- or underestimation of cost, each producing a similar failed result. An analyst faced with high technical uncertainty and wanting to be conservative may overestimate costs, excluding the application. An inexperienced analyst unaware of the technical risks may underestimate the cost with its subsequent consequences. One approach to dealing with technical risk is to contain it, rather than attempt to reduce it. Reducing the risk (e.g., changing the specifications to a proven technology) may also reduce or eliminate the associated benefits. Containing the risk means that development and implementation of the technology is tightly controlled with a known, affordable maximum expenditure. If failure occurs, the consequences were well understood and an appropriate course of action was known before hand. A commonly used technique is to build a prototype implementation with a fixed budget and predefined success criteria. Prototypes (discussed in Chapter 6) can be used in conjunction with pilot projects (to be discussed) to assess other aspects of the technologies' use.

A popular approach to dealing with uncertainty is to undertake a pilot project to test concepts and implementation issues. Pilot projects create temporary working environments in which costs, technology, staff, and organizational impact can all be controlled. They create an environment where failure (i.e., something not working as predicted) can be tolerated. However, for a pilot project to be successful, there should be specific measurable purposes. A pilot project should not be undertaken just because no one knows what to do next or because it seems like the thing to do. There are three primary reasons for undertaking a pilot project:

- *As an experiment,* the purpose of which is to investigate a particular technical aspect of implementation such as a data processing technique or to determine that a selected application is technically feasible and practical
- *As a demonstration,* during which potential users of the system are shown what is possible and the utility of a variety of demonstration applications; this can be combined with one or more educational seminars
- *As a temporary operation or production environment,* during which agency data are processed to answer specific problems, to assess operational feasibility, and to determine organizational impacts

Pilot projects can of course combine elements of each of these to satisfy a broader range of requirements. The common key element for a pilot project to be successful is for the sponsors and other participants to have a clear understanding of why the pilot project is being undertaken and what the measures of success are. In the absence of clear objectives, a pilot project may carry on indefinitely until sponsors and users lose interest. Similarly, a successful pilot project may never be deemed as successful simply due to the lack of performance measures that would establish its success.

A pilot project should be part of a larger implementation plan that includes the elements already discussed. A pilot project should fit within the context of the broader management strategy for introducing and managing change, determined during strategic planning.

Pilot projects can satisfy different purposes at different stages in implementation. In the early stages, a pilot project may be used as a demonstration to build awareness and interest. It can also be used to identify potential problems and sources of error. Conducting a pilot project can help to uncover these issues early enough in the process to deal with them cost effectively, but a pilot project is not and should not be used as a substitute for analysis and design (as discussed in Chapter 5). Later, during system construction, a series of well-planned and -constructed pilot projects can form the framework for incremental implementation of complex applications; this idea is developed in Chapter 6 in the section on managing pilot projects.

Implementing the Plan

Implementation can begin as soon as the startup plan has been prepared and agreed upon. It is worth noting that planning does not cease at this point, and that plans should not be cast in stone. However, some stability is necessary for any results to be achieved. The balance between flexibility (the ability to modify plans) and stability (having stable specifications from which to work) can in part be achieved by determining suitable planning horizons and time periods between review points.

No matter how carefully the planning has been done and even with a commitment to quality, some changes will still be required. End users in most organizations will have long learning curves, and substantive experience is not gained until they begin working with a system. As they learn, end users and technical staff will identify new opportunities and modifications to applications in progress or already implemented. One way to deal with this issue is to forbid changes. However, legitimate savings and improvements may be lost by following that course of action (aside from the fact that it goes against the participating high-commitment culture that we have been trying to build). At the other extreme, pursuing every new idea will deliver few useful results and will send costs through the roof. A *change order* process permits new ideas and changes to be recorded and analyzed prior to committing to them. It thus provides a mechanism for encouraging innovation without chaos. Change orders are assessed for impact and cost, assigned priority, and sent to the management authority for approval (this process will be expanded in Chapter 6). Such a change order process protects the implementation manager, maintains end-user enthusiasm, and provides the necessary management controls.

Many of these issues and others have little or nothing to do with GIS; like most other activities, they have to do with managing people. DeMarco and Lister (1987) make the point about the importance of dealing with human issues:

> The main reason we tend to focus on the technical rather than the human side of the work is not because it's more crucial, but because it's easier to do. Getting the new disk drive installed is positively trivial compared to figuring out why Horace is in a blue funk or why Susan is dissatisfied with the company after only a few months. Human interactions are complicated and never very crisp and clean in their effects, but they matter more than any other aspect of the work. If you find yourself concentrating on the technology rather than the sociology, you're like the vaudeville character who loses his keys on a dark street and looks for them on the adjacent street because, as he explains, "The light is better there." (p. 5)

Summary

In this chapter we have tried to highlight some the issues that in our experience come up most frequently. Still, the reader will likely encounter many others. There are rarely easy answers to dealing with people issues. However, creating a pleasant work environment, promoting open communication and participation, fostering trust and commitment, and rewarding enthusism and innovation will contribute much toward a successful implementation.

In summary, the following ingredients have been observed repeatedly in successful GIS implementations:

- *Planning*. Describe the overall concept for the GIS and subdivide that into manageable pieces. For each piece define what needs to be done, how long it will take, who will do it, what it will cost, what other resources will be needed, and who else needs to know about it. Write these up and give the paper to anyone who is affected by it. Statistically, most projects fail as a result of poor planning (Metzger, 1981). The initial project plan is the keystone on which the project's success or failure is based (Baumgartner, 1986).
- *Application driven*. All implementation activities from analysis to data conversion to buying equipment must be linked to an application linked to an end user who wants it. Applications are linked to business functions that provide the necessary services to the organization. If implementation is application driven, the opportunity for extraneous activity and expenditures is eliminated.

- *Support of the organization.* Ensure participation, and subsequently commitment, from all parts of the organization from sponsors (senior management), operations management, systems groups, end users, and any outside agencies. Committed, actively involved participants will want to make the implementation a success.
- *End-user participation.* End users (as the name implies) will be the folks most directly affected by the system. They are also the ones who best know what is being done now. End-user participation in analysis, design, pilot projects, tests, and other aspects of implementation will result in end users who are committed to success, understand why the system is as it is, and are willing make required changes to work procedures.
- *Measurable objectives.* Measurable objectives provide benchmarks against which to measure progress. They provide the means to assure project participants that the project is proceeding as planned or to provide the supporting evidence for needed changes to the plan when things are not.
- *Effective documentation.* Provides written records of agreements and objectives, commitment of funds or staff, forewarning of upcoming needs, background for new participants, reminders for long-term participants, and substantiation should past actions be challenged.

NOTES

1. For more information on managing technical people, the reader is directed to Weinburg (1986).

2. According to Alexander et al. (1977), "The importance of communal eating is clear in all human societies. Holy communion, wedding feasts, birthday parties, Christmas dinner, an Irish wake, the family evening meal are Western and Christian examples, but every society has its equivalents. There are almost no important human events or institutions which are not given their power to bind, their sacral character, by food and drink" (p. 697). The anthropological literature is full of references. For example, see Cohen (1961) and Richards (1932).

3. *Skunkworks* originated in the engineering profession. The term suggests that the project is hidden away from the rest of the organization, originally even from upper management. In a skunkworks project a team is given a goal and resources, and allowed to operate outside official channels. Tremendous energy and dedication can be developed by these teams. They tend also to be elitist and may often come into direct conflict with formal organizational authority.

4. JAD was pioneered by Arnie Lind while he was at IBM. He has now enhanced the method to bring the same productivity to many phases of the system design process (Lind Consulting, Inc., Regina, Saskatchewan, Canada).

5. John Rockart (1979) popularized the simple concept of critical success factors (CFSs)—typically, 6–10 factors viewed by executives as critical to meeting strategic goals. In government, these would be closely linked with mandates and legislation. Rockart found that few companies' information systems supported their CFSs. Most management information systems processed largely internal accounting-based and historical information. The significance of this finding to GIS is equally relevant, drawing

attention to higher-level decision-making and support those functions with GIS, rather than applying it solely to support functions such as mapping and record keeping (the GIS equivalent of accounting).

6. More information on PERT and related techniques can be obtained in Weist and Levy (1977).

7. According to DeMarco and Lister (1987), "the software industry has come to accept a defect density of one to three defects per hundred lines of code" (p. 21).

8. Scott (1987, pp. 212–15) found that (1) the greater the technical complexity, the greater the (organizational) structural complexity—differentiation of structure by function, level, or location; (2) the greater the technical uncertainty, the lower the formalization and centralization of decision-making; and (3) the greater the technical interdependence, the more resources must be devoted to coordination mechanisms—rules, schedules, and line and staff officials.

REFERENCES

Alexander, C., S. Askikawa, and M. Silverstein, with M. Jacobson, I. Fiksdahl-King, and S. Angel (1977). *A Pattern Language: Towns, Buildings, Construction.* Oxford University Press, New York.

Baumgartner, John S. (1986). *Project Management.* Irwin, Homewood, Ill.

Cohen, Yehudi A. (1961). Food and its vicissitudes: A cross-cultural study of sharing and nonsharing. In *Social Structure and Personality: A Casebook.* Holt, New York.

DeMarco, Tom, and Timothy Lister (1987). *Peopleware: Productive Projects and Teams.* Dorset House, New York.

Keen, Peter G. W. (1991). *Shaping the Future: Business Design Through Information Technology.* Harvard Business School Press, Boston, Mass.

Metzger, Philip (1981). *Managing a Programming Project,* 2nd ed. Prentice-Hall, Englewood Cliffs, N.J.

Richards, Audrey I. (1932). *Hunger and Work in a Savage Tribe: A Functional Study of Nutrition Among the Southern Bantu.* Free Press, Glencoe, Ill.

Rockart, John (1979). Chief executives define their own data needs. *Harvard Business Review,* March–April.

Scott, W. R. (1987). *Organizations: Rational, Natural, and Open Systems,* 2nd ed. Prentice-Hall, Englewood Cliffs, N.J.

Weinburg, Gereld (1986). *Becoming a Technical Leader: An Organic Problem Solving Approach.* Dorset House, New York.

Weist, Jerome D., and Ferdinand K. Levy (1977). *A Management Guide to PERT/CPM: With GERT/PDM/DCPM and Other Networks.* Prentice-Hall, Englewood Cliffs, N.J.

5

Systems Design Methodology

Il ne faut pas vendre la peau de l'ours avant de l'avoir tue.[1]

PETER CROSWELL

Once the vision for a geographic information system has been established and supported, there are two ways for the project to fail. First, the project team can compare the various commercial products available and decide which one is the "best." Second, the project team can assume that most commercial products are so similar that a comparison of capabilities is not necessary and then select one based upon price, popularity, or some other factor. In either case, the analysis ignores the specific organizational and institutional environment under which the participating users will operate. With over 60 different commercial products available in the marketplace today, each with its own unique capabilities and characteristics, the appropriate hardware and software for a GIS in one jurisdiction may not be appropriate for another jurisdiction. This is because each GIS project has its own unique combination of users participating in the project and each agency has its own particular needs, resources, and operating environment.

Building a GIS is a matter of constructing graphic and nongraphic databases, developing or obtaining information processing capabilities, installing the appropriate computer hardware and software, and then implementing the organizational, procedural, and staffing changes needed to operate and use the system successfully. These are the essential major tasks to be accomplished, but they cannot be started until all participating users know what it is that they expect the system to do for them. Without a clear definition of how each participant expects to use the system, it will be difficult, if not impossible, to build databases, select commercial products, and then use the system. Thus, as with most information systems projects, knowing how the system will be used forms the basis for determining what information will be stored in it and what additional technical, organizational, and legal resources will be needed to be able to use that information. This process is called *systems design*.

What Is Systems Design Methodology?

Simply put, systems design is a method for determining what changes must be made to an organization's current environment in order for an information system to achieve its purpose. Prior to implementing a GIS, potential users have their own way of performing work with resources and procedures that have been in place for a number of years. Maps and land-related data have been stored and maintained manually or in computer files designed for very specific purposes. A GIS usually changes all of that because it is an *information system,* focusing on a multipurpose database, rather than an automated mapping system focusing on computerizing a process.

Outside the conversion of geographic information from paper maps and manually stored attribute data, this process can be the most time-consuming task in the systems development life cycle of a GIS project because there are many different types of needs to consider and because there is more than one user involved, each with his or her own particular needs. Systems design determines the answers to many difficult questions:

1. What maps must be produced from the system?
2. What data must be available from the system?
3. How are the maps and data to be processed?
4. Who will update what data?
5. How will updates be disseminated to the users?
6. What hardware and software are needed?
7. Who needs what types of equipment?
8. What technical staff is needed to support the system?
9. Who needs to be trained, and what training is needed?

This list is not exhaustive. In deliberating the answers to these questions, other related questions will arise—questions regarding standards, procedural changes, organizational changes, and so on. The questions developed by the International Association of Assessing Officials (1989) prior to a 2-year study to prepare GIS guidelines for assessors is listed in the Appendix. These questions cover the topics of administration, training and education, hardware and software, database design, use and selection of consultants, vendor selection, and cooperative efforts.

Generally, the many questions that must be answered during systems design can be categorized into the following nine categories:

- Functions
- Data
- Applications
- Hardware and software
- Staffing
- Training

- Procedures
- Organizational and institutional changes
- Legal considerations

Addressing the needs in any one of these categories will assist in addressing the needs of other categories listed below it. For example, once the functions of an organization are identified, it is easy to define what data are needed to perform that function and how those data must be processed: To perform the function of "updating property maps," an agency must have the data recorded from a survey, and those data must be useful for placing new lines on the map. Similarly, once it is clear what the hardware and software needs are to support the data and processing needs of an organization, it is easier to identify staffing needs and then, in turn, the training needs. This hierarchy of needs is portrayed graphically in Figure 5.1.

Without exception, however, the functions of the organization determine the parameters for the changes needed in all other categories. This is because the functional responsibilities of each office within each participating agency define

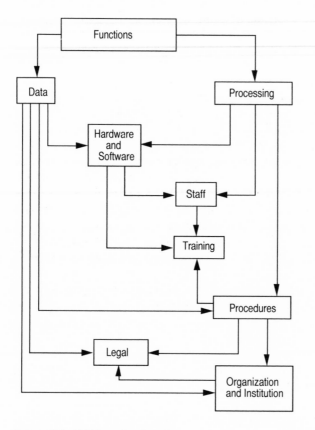

FIGURE 5.1 GIS hierarchy of needs.

how the system will be used. All other needs are then based upon these uses. For example, "updating maps" is one of the most common functional responsibilities of most offices contemplating the use of a GIS. Once the particular map is identified (e.g., property map) as well as the sources of information needed to update it (e.g., survey notes, subdivision plans, drawings), the specific *data* needs can be documented. The form of the sources of change also define the *applications* by defining how the information is to be processed (e.g., COGO[2] processing of survey notes, precision placement of subdivision plans). Without a clear understanding of all of the functions of the potential users, it is difficult, if not impossible, to define clearly all of the changes that must be effected so that the GIS will be successfully implemented and used. This functional approach to systems design ensures that the system will be an integral part of the operation of each agency rather than some research activity not directly related to their mission.

A functional approach to the process of collecting and analyzing the data requirements of an organization, especially a large organization, results in complex relationships between data and data usage. This problem is not unique to GIS implementation, and in recent years new tools that build upon structured methods have been developed. These tools are not in place of a structured design methodology, and in fact are only useful in combination with one.

Tools for Planning and Engineering the Design of Information Technology

System design tools have been used in the information systems field since the 1970s when system development methodologies such as *structured programming, structured design,* and *structured analysis* began to be used to improve the design and programming of computer application programs. These methodologies had two benefits: They provided rigorous procedures for designers to follow (to ensure that no errors or omissions were present in their programs), and they provided a standard method for documenting the design (so that when modifications were later necessary, others could easily understand the specifications and logic of the programs). The commercial sector has assisted in the use of these methodologies by developing computer programs that information systems professionals can use in this process. The result is a variety of commercial products called *computer-assisted software engineering* (CASE) tools for developing information systems.

While most CASE products were designed and used to automate the process of developing computer programming code rather than the higher-level information systems design process, many now integrate information systems planning and design with the automated coding feature to provide a comprehensive array of automated tools for assisting in strategic planning as well as design and implementation. However, there now appear to be two categories of CASE tools, each with a different emphasis:

1. Process flow, where the emphasis of design is on the logic used in the

computer processes. Thus, the term *software engineering* as described is commonly used to describe these products.

2. Data structure, where the emphasis is on the data stored and maintained in the systems. The term *information engineering* is commonly used to describe these products.

CASE products that provide strategic planning capabilities as well as design and program coding are available from both groups and appear to be equally used in the industry. However, those that emphasize information engineering are more commonly used in organizations that make extensive use of commercial software and fourth-generation languages, as opposed to those organizations that prefer to develop their own programming code. Since GIS technology almost exclusively consists of proprietary software developed by each GIS vendor, the information engineering approach to CASE tools is most appropriate for GIS projects.

Information Engineering

Information engineering tools have as their basic premise that the types of data used in an organization are quite stable but that the procedures, processes, and organizational structures that use the data are likely to change frequently. This is explained in Martin and McClure (1983):

> The methods and procedures and the information needs of executives change fundamentally. The computer programs, processes, networks, and hardware change, but the basic types of data remain the same. The data thus form a relatively stable foundation on which computer processes are built. These processes will change constantly and it is desirable to make this change as quick and easy as possible. (p. 299)

To be sure, data values change frequently as changes in the real world are observed and as databases are updated to reflect the latest conditions important to the organization. Data values in the tax rolls in local government, for example, are updated regularly to reflect changes in property owners and property characteristics for tax assessment and billing purposes, but the type of data in a tax roll database (owner name and address, parcel size, land use, etc.) remain constant from year to year.

This emphasis on the data needed by an organization allows information engineering tools first to develop a data model of the organization that meets the strategic information needs of an organization and second to identify whatever organizational or technical barriers are currently in existence that might prevent data sharing on an organization-wide basis. It treats data as a basic organizational resource that must be managed—just as the resources of people, equipment, and money are managed in an organization.

The results are information systems directly linked to organizational strategy and goals, satisfying organization-wide requirements, and providing for the most effective and efficient use of information in achieving the organization's mission (Arthur Young & Company, 1987). This is done by involving users directly in the development process and building logical models of business data and activities.

Arthur Young & Company (1987) identified seven objectives of information engineering in the development of organization-wide information systems:

- Responsive and accurate support to the information needs of senior managers by developing information systems that are of strategic importance to the organization
- Focus of the IS team on relating information system activities and products to the organization goals and critical success factors they support
- Providing senior management with an increased understanding of, and a greater ability to control, the organization's information systems resources
- Assistance to the organization in gaining and maintaining a competitive advantage in the marketplace by identifying strategic uses of information technology
- Decreasing the time required to bring new applications into productive use, and reducing the maintenance problems associated with keeping them cost effective and productive
- Involving users more effectively in information systems development through the increased use of techniques such as joint application development (JAD) and prototyping
- Improving the quality of information systems software by increasing the rigor of the methods used to create it, and by basing the system design on data and activity models of the underlying business

Information engineering generally consists of four major phases:

- *Information systems planning.* The development of a high-level model of the organization and its data. This model links information requirements to senior management's strategic planning and identifies the major business areas of the organization by using techniques such as clustering and factor analysis of processes and entities.
- *Business area analysis.* A logical model of each business area identified in the previous phase that reflects the interrelationships of data, activities, and organizational goals of the business area. A detailed data model and activity model for each business area is developed and compared with an assessment of the current business area environment for developing alternative design concepts. These design concepts are prioritized via an organization benefit analysis and a specific implementation approach is selected.
- *System design.* The transformation of the requirements of the business areas into detailed application system specifications. Prototyping is used

to involve users in specific product design and the completion of database design. This phase also designs the conversion system, system test plans, acceptance criteria, and user and operator training programs.
* *Construction.* The automatic generation of programming code and development of procedures for system operation.

As mentioned previously, the last phase, construction, most likely will not be undertaken in the planning and design of a geographic information system because programs are packaged into commercial GIS products. The system design phase instead produces the system specifications, which are then used to select a specific configuration of GIS hardware and software from available vendors.

This process can take months to a year or so and requires a significant investment in human resources to be effective. It can also be difficult to complete if top management is not actively involved or if there are extenuating factors that impede its logical progress. Huxhold (1993) cites some examples of such factors that can impede this process in a government setting:

1. Existing laws or mandates by different levels of government may have established subgoals for individual offices or departments. This can result in a misalignment between the individual and subunit goals with the goals of the organization as a whole because skills and motivation are focused on those subunit goals. The organization-wide goals, then, can become secondary to the subunit goals (Finch et al., 1976; Campbell, 1991). This happens in local government, for example, when department managers such as a municipal engineer, a tax assessor, and a county clerk report directly to elected officials.
2. "Parochial outlooks and limited objectives of each department and profession will interfere with a comprehensive, multiple use view of the organization's requirements" (Brown and Friedley, 1988, p. 24). This implies that a government may not have an organization strategy, or that more than one strategy exists because of external political or legislative pressures.
3. The multidisciplinary nature of GIS technology can make it difficult for potential users to agree on system requirements. Thus "coalitions are continuously being formed and reformed with each group trying to structure the available resources to serve its interests (Campbell, 1991, p. 3).

With strong top management involvement and leadership, however, these obstacles can be overcome and information technology planning can be successful in developing organization-wide geographic information systems. These techniques are applied in determining the relationships between data management and data usage, as described in the following section.

The Functional Approach to Systems Design

The functional approach to systems design determines how the use of the system will support the mission and responsibilities of each participant in the project. It is not a sophisticated process. It is merely a systematic method for ensuring that the use of the system will appropriately meet the needs of each user, and is facilitated through a series of information gathering and documentation tasks that produce a *functional business model*. This functional business model is a description of *what* it is that the organization does: its functions. It is not a description of *how* it works, as is documented in the traditional organization chart. Rather, it focuses on the functions performed by the organization and how data can support those functions.

By studying the organization from this functional aspect, rather than its current organizational structure, the information systems design will be assured to be comprehensive and integrated within the organization as a whole, rather than being constrained by existing programs, policies, and procedures. This is because the organization chart represents the current structure that is in place to perform the business functions. Change to the organization structure does not, by itself, require change in its information systems; however, change in the organization's business functions typically requires change to its data and information systems. A given activity, for example, may, at any time, be assigned to a different organizational unit; however, this in itself does not change the data and information system requirements of that activity. In addition, the organization chart represents the current organizational structure that may be based upon other, sometimes temporary, restrictions such as personnel changes, size of individual units, and personalities. The functional business model is thus more stable than the organization chart.

The value of using functional analysis as the basis for design is that it builds independence between organizational structure and information technology planning. This functional analysis can, however, result in a new organizational environment in which common or related functions are combined under a single office or otherwise changed in order to perform the work necessary to fulfill the mission of the organization more efficiently or effectively (Arthur Young & Company, 1987).

Functional analysis consists of three major elements:

- Documentation of current functions performed, resulting in a *current physical model* of the organization
- Development of a logical equivalent of the current physical model, resulting in a *current logical model* of the organization
- Application of management, legal, and other constraints to the logical equivalent, resulting in a *new logical model* of the organization

This process is similar to an operational audit of the organization—a process of asking why work is performed in the manner it is (the current physical model)

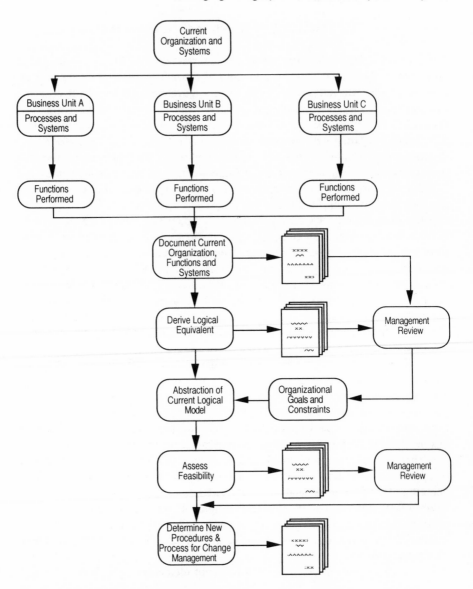

FIGURE 5.2 Functional analysis process.

and a determination of a better way that it could be done (the logical model), given existing constraints that have imposed by management directives or legal actions. Figure 5.2 depicts a schematic view of this process.

Current Physical Model

The current physical model of an organization is a comprehensive list of the work performed by all operating units of the organization. This work, also known as functions, consists of the activities an organization performs to carry out its mission. They are processes involving human activity (such as moving, thinking, speaking, and decision-making) that require information and that can be improved by improving the quality or method for obtaining or processing the information. These activities occur at all levels of the organization, from delivering products or services, to managing resources, to setting policy and long-term strategies. Improving the quality and access to geographic information can improve many of these activities; so the functions that agencies perform must be reviewed in order to identify what maps and data are needed and how they are processed.

All participating users of a GIS can identify what functions they perform from a number of different internal documents such as organization charts, mission statements, annual budgets, and long-range plans. It is important to review all of the activities of the agency because these analyses often focus only on maps and data displayed on maps; there are also many functions that rely on information related to locations, but not necessarily on maps. For example, the County Sheriff's office may be required to "process subpoenas"—to deliver legal notices to people ordered to appear in court. While this function may not require a map (other than to find an obscure address), it does require geographic information: The subpoenas may first be distributed to squads assigned to subareas of the jurisdiction before serving them, or they may be served in a geographic sequence that minimizes the travel time between the addresses of the people being served. In this case, a GIS can improve the process of serving subpoenas even if the process does not require a map.

The initial step in defining functions is to identify all of the organizational units (departments, divisions, bureaus, offices) and list the functions they perform that require maps or other geographic information. This need not be a tedious or time-consuming task—especially if existing budget or mission documents are readily available—and can be accomplished by one individual. It is important, however, not to omit any function that may conceivably benefit from a GIS (such as "process subpoenas"); later on (i.e., after implementation) it may be simple to enhance the system to benefit that function if the appropriate standards and resources are established early in the project. Furthermore, what may seem like a trivial application to the project team may actually be of high importance to the nontechnical end user who sees a tremendous benefit from the application and, thereby, provide some much-needed success early in the project.[3] Table 5.1 presents a sample list of functions requiring maps or geographic information by organizational unit within a local government.

The current physical model, then, is the documentation of what functions are performed in an organization and which functional units perform them. The model may be documented in any fashion that communicates this information: lists of functions by organizational unit (as depicted in Figure 5.2); a matrix of cells showing which functions are performed by which organizational units (Figure 5.6); or any other mechanism that is effective in relating the current operation and responsibilities of the organization.

Table 5.1   Sample List of Local Government Departments and Their Functions

Department	Function
Board of zoning appeals	Hear and decide appeals
Building inspection	Inspect buildings Issue code violation notices Review building plans Issue permits
Capital improvements	Prepare CIP budget
City clerk	Issue licenses Enforce license regulations Research current issues
City development	Perform development research Analyze trends Upgrade public housing stock Assist community groups Manage public land Dispose of surplus property Review development proposals Review/prepare zoning changes
Common council	Provide political leadership Discharge legal responsibilities Respond to citizen complaints
Election commission	Maintain voter registrations Establish voting wards
Fire services	Respond to fires Respond to medical emergencies Plan for fire resources
Harbor commission	Market port facilities Acquire waterside facilities
Health	Process vital records Examine infants/children, others Monitor TB, VD, AIDS, other Investigate environmental problems Inspect food establishments Investigate toxic/hazardous material
Mayor's office	Coordinate long-range planning Handle citizen inquiries/complaints Coordinate research and policy

Department	Function
Police services	Respond to calls for service
	Manage police resources
	Organize block clubs
	Analyze crime trends
	Investigate crimes
	Investigate traffic accidents
	Patrol neighborhoods
	Enforce traffic laws
	Apprehend/arrest criminals
Public works	Issue permits in public way
	Prepare special assessments
	Prepare paving program
	Construct/maintain bridges
	Plan/design sewers and laterals
	Investigate flooding complaints
	Prepare/maintain maps
	Plan/design roadways
	Review/process developments
	Provide engineering services
	Inspect infrastructure construction
	Evaluate/plan pavements
	Repair/replace public trees
	Maintain boulevards, parks, other
	Dispatch municipal vehicles
	Plow snow/salt ice
	Collect/dispose of solid waste
	Sweep streets and alleys
	Maintain sewers/drainage channels
	Repair/replace pavement/sidewalks
	Design/repair traffic facilities
	Design/repair street lights
	Design/repair municipal communications
	Install/maintain lane markings
	Read water meters
	Maintain water facilities
Tax commission	Appraise properties
	Prepare assessments
	Perform title searches
	Maintain parcel maps
Telecommunications	Monitor/inspect cable TV installation
Treasurer	Bill/collect property taxes

Current Logical Model

The way work is currently being performed as documented in the current physical model is not necessarily the way work should be performed in the organization. Over time, as the organization grows and accepts (or is mandated to accept) new responsibilities, and as others are eliminated, changed, or turned over to the private sector, the organizational responsibilities change as functional responsibilities are moved, added, or eliminated. These organizational changes may not be enacted comprehensively, with a thorough analysis of all of the interacting processes among the various operating units within the organization. While the current physical model documents how the functional responsibilities are currently aligned within the organizational structure, the functional analysis process of GIS needs assessment allows top management to review the efficiency and effectiveness of these alignments and identify potential changes for improvement. This is done by developing a logical equivalent of the current physical model through a process called *functional decomposition*.

The functional decomposition process asks the question: "Are current functions assigned to the appropriate organizational units?" It uses the current functions identified in the current physical model as given (they must be done to meet the mission of the organization), but does not accept the current responsibilities for the functions as given. It groups functions logically rather than physically. Take, for example, all the map updating functions given in Table 5.1. In this list of functions assigned to departments in local government, four different departments are listed as having nine functions related to updating maps used in local government:

City development	Review/prepare zoning changes
Election commission	Establish voting wards
Public works	Plan/design sewers and laterals
	Prepare/maintain maps
	Plan/design roadways
	Design/repair traffic facilities
	Design/repair street lights
	Design/repair municipal communications
Tax commission	Maintain parcel maps

A logical view of these functions is to assign all functions that involve the updating of maps to one office. (Based upon the functions listed, it might be logical to have one organizational unit update all maps: zoning, voting, sewer, roadway, traffic, street lights, communications, and parcels.) Thus, the logical equivalent of the physical function "map updating" is recorded in the current logical model for all activities in the organization that require a map to be updated, regardless of the particular map requiring updates. Logically, then, one might assume that all functions of an organization that require a map to be updated should be performed by the same organizational unit (in order that the drafting expertise can be concentrated, resulting in the efficient utilization of skills and resources). In this example, then, the functional decomposition might eliminate all of the functions related to map maintenance in various departments and bureaus and leave one function in the public works department for all of those activities: "prepare and maintain maps."

The functional decomposition process, then, decomposes the functions of the organization from their current organizational assignments (as recorded in the current physical model) and then logically groups common functions to produce a current "logical" model of the organization.

New Logical Model

In the real world, however, logic does not always prevail. Both internal and external constraints imposed by management and policy-makers will prevent the logical grouping of work as proposed in the logical model of the organization. Internal constraints may be as simple as the physical location of people or documents or equipment that can prevent a function from being performed.

Take, for example, the source of the information that identifies when a building has been built and how that information can be recorded on a map that identifies building footprints. Generally, local ordinances require the issuance of a building permit before construction of the building can proceed. The authority to issue a building permit may lie within an organizational unit with the expertise needed to identify potential hazards in its construction. Some of this information will also be needed by the tax assessment function to assign a new assessed value to the property. If the building footprint map updating function is located in a third department or office, it may be inefficient to route the appropriate information to all three functional units.

Internal constraints on the logical assignment of functions may also be as complex as political compromises and coalitions that can be formed in government organizations. Police and fire departments may insist on complete control over the street segment database because emergency-response systems are the responsibility of these departments—in spite of the universal use of the street segment file for geocoding address-based records maintained by virtually every other department in the government. If these other departments are to realize the benefits of a street segment file for geocoding their records, they may be compelled to allow the police department to maintain the data (or they find themselves duplicating map maintenance functions).

External constraints (legal, economic, technical) can also throw logic aside and affect the assignment of work. Federal, state, provincial, and regional mandates can dictate the assignment of work in local government.

The new logical model of the organization combines the current required functions of the government, the logical grouping of these functions, and the reality of the current internal and external environment to produce a new organizational responsibility chart of the organization in order to improve its efficiency and the effectiveness. These functions must then be supported by the technology of geographic information systems by next defining the data needs and the processing needs of these functions.

Data Requirements Definition

Determining what data are needed for processing in a GIS is more than merely listing the data items that potential users say they want to be stored in the computer. Since the cost of converting data to digital form (data conversion) exceeds by far the cost of any other component of the system (between 50 and 80 percent of the total cost of the project), it is important to verify that each data item is, in fact, essential (Antenucci, 1990; Huxhold, 1991). What is the purpose of storing elevation contour lines in the system when field surveys are taken each time a construction project is undertaken? In addition, each item stored in a GIS must be kept up to date during the continued operation of the system and there is a cost associated with that "data maintenance." (For example, if building footprints are needed, there is a cost associated with updating that information as buildings are built, modified, and demolished.) Thus, the determination of the data items needed in a GIS must be more than a "wish list" agreed upon by all users—it must involve a systematic study of how valuable each data item is for the functions of each user. Later in the needs assessment process, it will be important to determine procedures and data maintenance responsibilities required to keep the information up to date once it has been converted to digital form.

Geographic Information Needs Inventory

A more systematic and comprehensive method for determining data needs are to take an inventory of the data that are currently being used; if they are being used now, they will most likely be used once they are converted to digital form. This can be accomplished by surveying each functional unit within the organization to identify what map and other geographic data are used in each function it performs. This process will not only provide a comprehensive approach to needs assessment (ensuring that all functions of an organization are reviewed), but it also ensures that the data that will eventually reside in the GIS will also be used on a continual basis.

The inventory of geographic information needs can be accomplished by the use of a survey document such as the example shown in Figure 5.3. This survey document, sent to each functional unit of each participating user, should be completed by the person responsible for the functions of the unit. It is indexed first by the primary agency, department, or bureau and then by the functional unit within the agency, department, or bureau (much the same as the index of functions from Table 5.1), and includes a brief description of the mission of that unit. In many cases, each functional unit may include a number of smaller subunits such as divisions or offices—each having separate and distinct

AGENCY: _____

FUNCTIONAL UNIT: _____

    MISSION: _____
                  _____
                  _____
                  _____

    SUB:UNIT: _____

Functions	Maps or Drawings Used	Maps or Drawings Produced	Other Data Used	Source
_____	_____	_____	_____	_____
_____	_____	_____	_____	_____
_____	_____	_____	_____	_____
_____	_____	_____	_____	_____
_____	_____	_____	_____	_____
_____	_____	_____	_____	_____

Current Map or Data Problems:

_____
_____
_____
_____
_____
_____
_____

Future Needs:

_____
_____
_____

FIGURE 5.3 Geographic information needs survey form.

functional responsibilities.[4] In these cases, it is advantageous also to subdivide the inventory by having each subunit complete its own survey document.

Completion of the geographic information needs inventory survey document will identify for each function of the unit or subunit which maps or drawings are important for successful completion of that function. It also identifies what other data are used in the function if they are not recorded on the map or drawing. Recording this information for all participating users is a critical step towards the final determination of the content of the databases required for the GIS. It also lays the foundation for the procedures that will be required for operation of the system because it identifies the *flow* of the maps, drawings, and data through the user offices (and, eventually, between offices). It records which maps, drawings, and data are used (received from outside the unit or retrieved from internal storage) and which are produced (or updated) for use in another unit or storage facility. This information—what geographic data are needed for a function and where they are sent after completion of the function—can be seen graphically in Figure 5.4, using the example of a permit review process.

The permit review function shown in Figure 5.4 is used as an example of the geographic information needs inventory. It is a common function of many local governments and utility companies, and it usually involves the use of many different types of geographic information flowing among many users—each reviewing a particular aspect of a proposed change in the physical environment of the jurisdiction. In this function, a developer submits a request (permit application form) for a change (development plan) to one unit of an agency for approval. This unit then retrieves (from storage) additional information

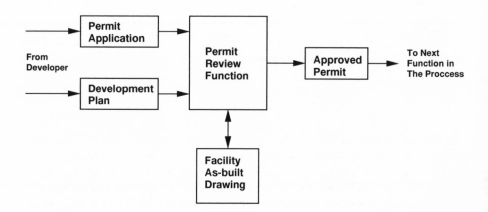

FIGURE 5.4 The flow of geographic information through a function: permit review example.

describing the facilities that are already in place (and where they are) in the form of an as-built drawing. Based upon the plan and the existing facilities from the as-built drawing, the review process may result in an approval and then transmission of the information to another unit (or another agency) where the change is reviewed with respect to different facilities. The geographic information needs inventory keeps a record of *what maps and other data are needed* (permit application, development plan, and as-built drawing) to perform the function (permit review), *where they come from* (the developer), and *where the results of the function are sent* (the next function in the review process).

A disadvantage of the inventory approach to data needs analysis, however, is that it addresses only the current situation (the data currently used and the procedures currently in place to obtain them and process them manually) and does not necessarily address improvements that can be achieved through automation. In some cases, data may be used in a certain function only because other, more accurate, or pertinent data are not available. In other cases, important data may not be used at all because it would be too time consuming and costly to obtain the data. Often estimates are made or factors ignored when having the extra data could improve the effectiveness of the function. Take, for example, a zoning map that does not include parcel boundaries on it. In many jurisdictions, zoning maps are separate maps with zoning district boundaries identified in relation to the public right-of-way and are maintained by one unit of local government. Since parcel boundary changes may be recorded by a different unit or even a different level of government, ongoing maintenance of parcel boundaries on the zoning map would be too costly—it would cause duplicative work by two different units of government. When zoning issues arise that affect properties near the boundary of zoning districts, the zoning maps are useless in accurately identifying each affected parcel.

To avoid automating an already inefficient or ineffective process, the geographic information needs inventory must also record information about problems or inefficiencies in the current use of geographic information. Later, in the design of the databases and processing functions of the system, these problems can be addressed and, possibly, corrected.

Another important consideration in completing the geographic information needs inventory is an anticipation of changes to the function of the unit in the future that may have an impact on the data needed to perform the function. These *future needs* may be as extreme as a reorganization of the agency or a new law that is likely to be enacted (such as environmental constraints), or it may be more subtle, such as estimates of increased volume of work (such as could be experienced in a growing jurisdiction). It is often advantageous for the GIS project team to assist the unit manager in completing the survey document in the form of an interview in order to raise issues about problems and future changes that should be considered in the design of the system.

Map Inventory

One of the results of the geographic information needs inventory is the identification of all maps and drawings used by the potential participating users. This inventory of mapping needs provides the background for the next step in the analysis of data needs: a detailed description of each map and drawing in the inventory (Figure 5.5). Since each map or drawing used in the functions of the participating users is a candidate for inclusion in the GIS, it is important at this early planning stage to have a detailed description of what features are included on them, what their geographic coverage, scale, and accuracy are, how they are updated, and what other functions use them. This inventory will be useful for

Map or Drawing Name: _____

Functional Unit Responsible
for Maintenance/Dissemination: _____

Geographic Coverage:
    Each Map/Drawing Sheet: _____
    Entire Set or Series: _____

Scale: _____

Accuracy: _____

Source of Base Information: _____

Causes for Updates/Changes: _____

Frequency and Volume of
    Updates and Changes: _____
Time Spent in Updating: _____

Sources of Updates/Changes: _____

Dissemination: Recipient   Frequency
                  _____   _____
                  _____   _____
                  _____   _____
                  _____   _____

List of Features Displayed: _____
                            _____
                            _____

FIGURE 5.5 Map inventory form.

- Determining what features need to be converted to digital form
- Estimating the amount of work (and potential cost) needed to convert the data
- Estimating the potential benefits that can be realized in reducing the time it take to update, create, and use them once they have been converted to digital form
- Determining the amount of redundant map maintenance that can be eliminated by combining the geographic data into one common database
- Assisting in the evaluation of hardware needs for storing and processing the data

The Information Needs Matrix

Completion of the geographic information needs inventory and the map inventory surveys provides the project team with a vast amount of information about how each participant uses geographic information, what maps are currently used, and what problems, inefficiencies, and future changes are anticipated in the use of the maps and data. Indeed, with each user having many different functional units and subunits, the volume of recorded survey results can be difficult to manage and close to impossible to digest for use in planning the project.

A systematic method for evaluating the results of the surveys is needed in order to apply the appropriate perspective to the remaining steps of the needs assessment. Specifically, it is important to be able to apply some degree of importance to the information obtained because it is unlikely that a GIS can be implemented to satisfy all the needs documented—at least in the initial implementation phase. It is also valuable to differentiate between the common needs of all participants and the unique needs that can be satisfied locally by individual users.

One method for summarizing the results of these surveys is to prepare a geographic information needs matrix that graphically displays the interrelationships and common needs of all relevant functions potentially affected by the system. The geographic information needs matrix lists all functions surveyed on one axis and all maps used on the other axis (Figure 5.6). The cells where each row and column intersect give an indication of the importance of the map in that function. If a map is not used at all in the function, for example, the cell is left blank. If a map is essential for that function, then the cell is fully shaded. Ancillary uses (ones that may not be important or are only important at certain times) can be shaded some gradient between these two extremes.

The geographic information needs matrix provides a management overview of the extent of common geographic information needs across all functions (represented by the maps having the most cells shaded) and the degree of dependency each functional unit has on geographic information (represented by the functions having the most cells shaded). This can be helpful both in setting priorities for applications and for determining the contents of a common digital land base for use by all participating users.

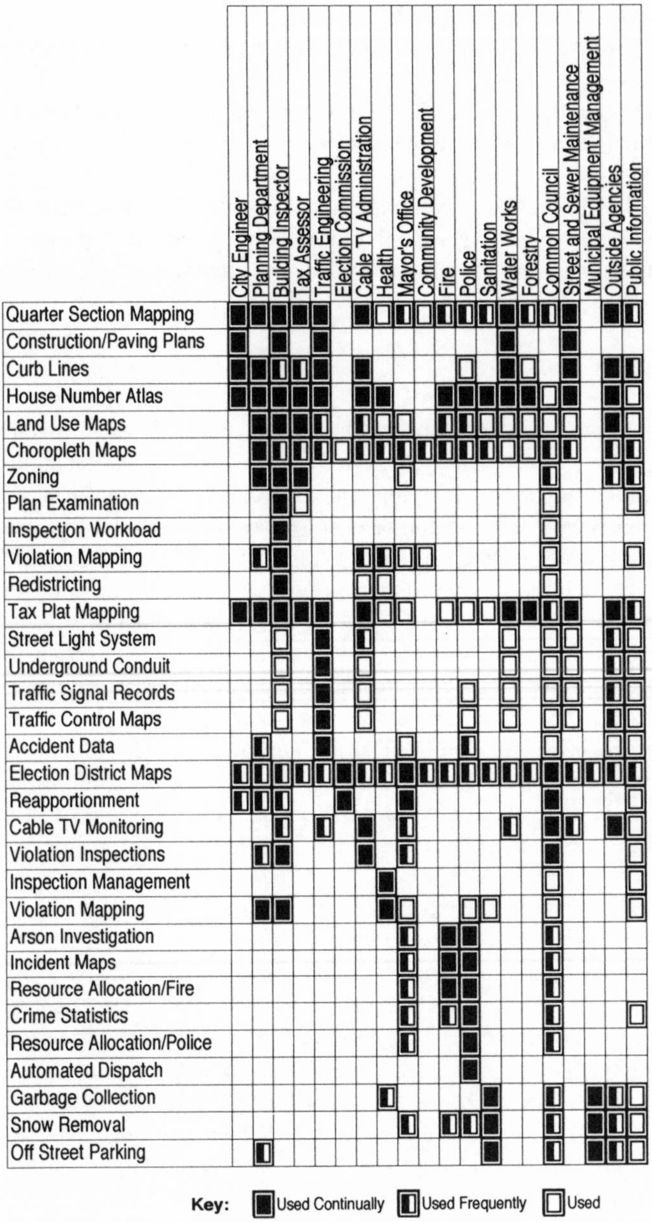

	City Engineer	Planning Department	Building Inspector	Tax Assessor	Traffic Engineering	Election Commission	Cable TV Administration	Health	Mayor's Office	Community Development	Fire	Police	Sanitation	Water Works	Forestry	Common Council	Street and Sewer Maintenance	Municipal Equipment Management	Outside Agencies	Public Information
Quarter Section Mapping	■	■	■	■			■	▣	□	▣	□	▣	□	▣	□		▣	■		▣ □
Construction/Paving Plans	■	■	▣											▣			■			
Curb Lines	■	■	□	▣	■		■					□		▣	□				▣	□
House Number Atlas	■	■	■	■			▣	▣		■	■	■	■	■	□		▣		■	□
Land Use Maps		■	■		▣		□	□	□		■	□	□	□	□	□	▣			■
Choropleth Maps		■	□	▣	□	▣	□	▣	□	▣	□	▣	□	▣	□	▣	□	▣	□	▣ □
Zoning		■	■					□								▣			▣	□
Plan Examination		■	□													□				□
Inspection Workload		■														□				
Violation Mapping	▣	■					▣	▣	□	□						□				□
Redistricting		■					□	□								□				
Tax Plat Mapping	■	■	■	▣	■		■	□	□		□	□	▣	□	■	▣			▣	▣
Street Light System		□	■		▣							□		▣	□				▣	□
Underground Conduit		□	■		□							□		▣	□				▣	□
Traffic Signal Records		□	■		▣						□		□		▣	□			▣	□
Traffic Control Maps		□	■		▣					□		□		□		▣	□		▣	□
Accident Data	▣			■				□			■					▣			▣	□
Election District Maps	▣	□	▣	□	▣	□	■	□	▣	□	■	□	▣	□	▣	□	▣	□	▣	□ ▣ □
Reapportionment	▣	□	▣				■		■							□				
Cable TV Monitoring		▣	□	▣		■		■					▣		▣	▣		■		
Violation Inspections	▣	■					■		▣							□				□
Inspection Management							■									□				□
Violation Mapping	▣	■					■	□		□	□					□				□
Arson Investigation								□		■	▣					■				
Incident Maps								□		■	▣					■				
Resource Allocation/Fire								□		■	▣					■				
Crime Statistics								□		■	□	▣				■				□
Resource Allocation/Police								□		■	▣					■				
Automated Dispatch										■										
Garbage Collection							▣						■			▣		■	▣ □	
Snow Removal									▣		▣ ▣ □	■			▣		■	▣ □		
Off Street Parking	▣											■				▣		■	▣ □	

Key: ■ Used Continually ▣ Used Frequently □ Used

FIGURE 5.6 Geographic information needs matrix.

The Shared Land Base. It is not necessary for a GIS to be a single configuration of hardware and software used by all participants because functions vary from agency to agency and different technology satisfies different needs. The integrating factor, however, among the participants of a GIS is the *information* contained in the system—the geographic data used by all participants to fulfill the requirements of their missions. Much of the information needed by the participants is common to all: street names, waterbody boundaries, building addresses, municipality and county boundaries, and so on. For each participant to create and maintain the digital records of this common geographic information independently is an inefficient use of resources and a potential problem with data accuracy and currency. Indeed, the major advantage of a GIS is the elimination of redundant map data maintenance because certain features that can be mapped are needed by all participants. Thus, a common database with features updated only by the functional unit that is responsible for the records can be of significant value to the other participants who need the data. This common base of geographic data is usually referred to as a *shared land base or a base map*.

By reviewing the geographic information needs matrix, it is possible to identify the sources of data for the shared land base. The maps that are used by all participants (those that have shading in at least one cell for each functional unit) are the ones that contain features that are potential candidates for a shared land base. By reviewing the features contained on these maps from their descriptions in the map inventory, it is possible to determine what the components of the shared land base should be.

Table 5.2, for example, lists the components of a shared land base for a typical GIS consisting of local governments and public utility companies. It is divided into four classes of features: survey control features, planimetric features, topographic features, and cadastral features. While the GIS applications of each participant may contain additional data that are crucial to their mission, it is these common data, generally of value to all participants, that are contained in the shared land base and used by all.

SURVEY CONTROL FEATURES. In order to ensure that all features are accurately mapped in relation to each other and to the earth, it is necessary to establish and record certain features that provide common geographic references on a continuous coordinate system for the geographic area contained in an GIS. This can be done through the use of *monuments* that are physically placed in the earth (for reference in future surveys) and whose locations are recorded in the shared land base (usually by state plane coordinates, universal transverse mercator coordinates, or latitude/longitude). In most areas of the United States, the Public Land Survey System (PLSS) Section and Quarter-Section monuments are used; however, other monuments such as the National Geodetic Reference System (NGRS) control stations and other locally defined monuments can also be used. Determination of the accuracy, density, and distribution of these monuments and the quality of their geodetic reference are significant issues in the process of agreeing on a common reference system in a shared land base.

PLANIMETRIC FEATURES. Physical objects that can be seen on the ground (roads, railroads, rivers, shoreline, buildings) are planimetric features that all partici-

Table 5.2   Components of a Shared Land Base

SURVEY CONTROL FEATURES
  MONUMENTS
    Section corners
    Quarter section corners
    Benchmarks
    Survey control (bearing &
      distance)

PLANIMETRIC FEATURES
  ROADS
    Edge of Travelway
    Centerlines
    Street Name
    Private Roads
    Medians/Boulevards
  RAILROADS
    Dimensions
    Railroad Name
    Railroad Right-of-Way
    Railroad Centerlines
  HYDROGRAPHY
    Wetland Boundaries
    Waterbody Boundaries
    Waterbody Name
    Floodway Boundaries
  MISCELLANEOUS
    Feature Name
    Airports
    Bridges
    Building Footprints
    Culverts
    Dams
    Driveways
    Fencelines
    Parks/Recreation Areas
    Piers
    Dimensions
    Retaining Walls
    Trees
    Walkways
    Wooded Areas

TOPOGRAPHIC FEATURES
  2' Contour Lines
  Sport Elevations

CADASTRAL FEATURES
  COUNTY
    County Boundaries
    County Name
  MUNICIPALITY
    Municipal Boundaries
    Municipality Name
  TOWNSHIP/RANGE
    Township/Range Lines
    Township/Range Name
  SECTION
    Section Lines
    Section Dimensions
    Section Number
  QUARTER SECTION
    Quarter Section Lines
    Quarter Section
    Quarter Section Number
  SUBDIVISION
    Subdivision Boundaries
    Subdivision Dimensions
    Subdivision Name
  CERTIFIED SURVEY MAP
    Certified Survey Map
      Boundaries
    Certified Survey Map
      Dimensions
  LANDS
    Land Divisions
    Land Division Dimensions
  LOT/PARCEL
    Lot Lines
    Lot Number
    Block Number
    Parcel Ownership Lines
    Parcel Ownership
    Parcel Address
    Parcel Number
    Parcel Owner
  EASEMENTS
    Easement Lines
    Easement Dimensions
    Easement Name/Purpose

pants in a GIS project use when recording information about their facilities and other location-related features. In many cases, these planimetric features are themselves monuments from which measurements are made when recording the locations of other features—especially those that are under or above the ground. For example, most water mains, gas mains, and underground electrical lines are located by perpendicular measurements from curb lines, edge of pavement, or street centerlines. Surveyors often use these planimetric features as benchmarks for spot elevation measurements. This means that the content and the accuracy of the locations of the planimetric features in a shared land base are important for two reasons: the management of the physical objects themselves (street pavement, railroad tracks, rivers, parks, bridges) and the use of their locations as a reference to locate other objects that may not be included in a shared land base (water mains, gas mains). For these reasons, it is important for the locations of planimetric features of a shared land base to be referenced to the survey control features described.

TOPOGRAPHIC FEATURES. When all or most of the participants of a GIS require information about the terrain in terms of its slope and elevation above sea level (hypsography), then the topography of the jurisdiction must be recorded and maintained in the shared land base. Commonly recorded topographic features include spot elevation values and their locations and contour lines having a vertical elevation interval of from one to five feet. This information is valuable to participants in a GIS who need to model gravity-based network systems (such as sewers and river basins) and also for estimating earth movement requirements for new developments and travelways (cut-and-fill analyses, site preparation costs). In mountainous regions, topographic features can be critical to the successful use of a GIS. Spot elevations, as described, are usually located at some physical feature (the corner of a building, the end of a bridge, a survey control monument). Contour lines, on the other hand, are much more difficult (and costly) to locate because they are usually created through the use of stereo digital orthophotography.

CADASTRAL FEATURES. Cadastral features are geographic features that cannot be seen on the ground (other than when planimetric features are also located where they are recorded such as fence lines and property lines). In their broadest sense, cadastral features represent the locations of legally defined boundaries (county and municipality boundaries, subdivisions and lot and block numbers, ownership parcels, easements). In their narrowest sense, cadastral records are those records that identify certain legal interests in land (National Research Council Panel on a Multipurpose Cadastre, 1980). However they are defined, cadastral features are important to most participants in a GIS because they provide a legal and administrative context for relating their facilities to the real world (the owner of a parcel of land, the governmental jurisdiction in which a parcel or facility is located, the address of a building). Parcel boundaries, easements, and other location-related data are generally government-regulated geographic features needed by many different functions of the participants.

The contents of the shared land base differ from one project to the next, depending upon the participants in the project and the unique needs of each jurisdiction. Two general factors, however, have a major impact on what will actually go into the shared land base: the cost of obtaining the information and the cost (and responsibility for) maintaining the information after it has been converted to digital form. Each item planned for inclusion in the shared land base must be discussed among all participants in terms of these two factors—what will it cost and who is willing to keep it up to date. Survey control information, for example, can be obtained by an expensive remonumentation effort (if needed) in which qualified surveyors reestablish the physical monuments in the field and these monuments are targeted for capture in aerial photography. Over time, as physical changes to the land occur, it must be the responsibility of some participant to ensure that the monuments remain intact and that necessary changes are recorded on the shared land base. Similarly, the planimetric features can be obtained from aerial photography, but the benefit of having the digital record of each type of feature must be weighed against the cost of obtaining it and maintaining it over time (once the aerial photograph is taken, all changes to the features—new and demolished buildings, new and vacated streets, and so on—must be recorded on the shared land base by the functional units responsible for approving or creating the changes). Cadastral features, unavailable through aerial photography (because they cannot be seen), must be registered to the survey control network, and differences with the planimetric features (fence lines, hedges, buildings) must be resolved. Changes after their digital conversion (real property boundary changes, new subdivisions, new easements, vacated public right-of-way) must be recorded on the shared land base by each participant who governs the changes. Those responsible for a GIS, therefore, must consider not only the cost and effort to create a common digital record of all features needed by the participants, but it must also identify and make provisions for the continued updating of those features once the shared land base is built.

Applications Requirements Definition

Applications of a GIS define how the data are to be used to fulfill the functional needs of the organization once the system is operational and can be identified by analyzing the functions of the organization that will benefit from the use of the system (see Table 5.1). Associating the applications to be developed for the system with the *functional needs* of the organization will ensure that the processing capabilities of the system are consistent with the overall mission and responsibilities of the organization.

Defining these applications can be initiated early in the project planning phase and often is accomplished during the analysis of data needs. This is because the applications require data for processing (inputs), and often produce products (outputs) that are used in the function or by other functions. For example, the permit review function depicted in Figure 5.4 requires three inputs (permit application, development plan, and a facility as-built drawing) and, upon completion of the review function (or process), produces one output (approved permit). This procedure—receiving inputs, processing the information from those inputs, and producing an output or some other result—forms the basic structure around which the processing needs can be identified and documented in order to define the applications that are needed in the system, as shown in Figure 5.7.

Determining GIS applications, then, is a matter of investigating each function of the participants and answering the following three questions for each:

1. What data are processed (inputs)?
2. How are the data used (application)?
3. What is done with the data after they have been processed (outputs)?

Once the functions that will be supported by the GIS have been defined (as recorded in Table 5.1), the applications that will be needed for processing the data contained in the system can be determined and documented. The geographic information needs survey form (see Figure 5.3) contains most of the basic information needed to define the application: functions, maps or drawings used and produced, and other data used. All that is needed is a little more structure to the analysis. This structure takes the form of identifying the step-by-step procedures used to complete each function (usually accomplished by interviewing the people involved or by referring to documented standard procedures).

Take, for example, the local government function identified in Table 5.1 as "Review/prepare zoning changes." In analyzing the processing needs of this function, it is possible to identify all of the steps taken by the functional unit to "Review/prepare zoning changes." A typical set of steps may include the following:

1. Receive zoning change request.
2. Review existing zoning in the vicinity.
3. Identify current land use for the area and surrounding area.
4. Compare the situation with established legal restrictions and requirements.
5. Identify and notify property owners in the area.
6. Conduct a public hearing on the request.
7. Prepare map and ordinance changes.
8. Obtain the necessary approvals.

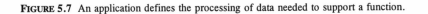

FIGURE 5.7 An application defines the processing of data needed to support a function.

Decomposing a function into these smaller steps makes it easier to identify the specific applications that must be developed for the system to assist in the processing of data in the function. For example, the second step, "Review existing zoning in the vicinity," requires a process whereby the system can be requested to display a map of existing zoning within a certain geographic area. Thus, the need to process data (review existing data) defines the application: "Display zoning map information for an area defined by the user."

In order to develop this application, however, the system must know two things:

1. What is the geographic area?
2. What, besides zoning, should be displayed?

The answer to the first question defines the data inputs needed for the application and the answer to the second defines the output product. Hence, once the application has been identified, it is necessary to identify the inputs needed to be processed and the outputs necessary for the function.

Data Inputs

The data that are input to an application come from two sources: data that are input externally by the user and data that are stored within the database. There is no other source of data for the computer (other than, perhaps, data stored within the program logic itself such as a rate, measurement, or some other parameter that seldom changes). Thus, the applications developed to satisfy the processing needs of a GIS must know what data are required to be input by the user and what data are required from the database.

In our zoning example, the data input required from the user is an identification of the geographic area that must be displayed. This can be accomplished in one of two ways: either the user defines the parcel of land (by an address, a parcel identifier) and the application determines how large an area to display (through a parameter stored in the program logic), or the user defines the area to be displayed (by first identifying the parcel and then defining the size of the surrounding area to display).

Data Outputs

Data can be output from an application in a number of different forms:

- Hardcopy maps
- Hardcopy tabular reports
- Screen map display
- Screen tabular display
- Digital file to be used in another application

- Image on microfilm, video disk, and such
- Database (when existing data are updated)

It is necessary to define the specific output medium required for each application so that the appropriate design methods for that medium can be applied. The most important consideration, however, in defining the data output from an application is what data must be output.

Our zoning example requires a map of zoning information to be displayed (presumably on a screen map display and probably also on a hardcopy map). It is obvious, however, that more data than just "zoning" data are required on the output. What is not obvious is just what that other data are—parcel boundaries, parcel dimensions, street names, owner names, addresses, water main sizes. Just what data must be displayed? An explicit definition of each data item needed to be output as well as the medium upon which they are to be output is needed.

Documenting the Applications

In order to keep a record of all the applications desired for a GIS, it is helpful to maintain a standard format that identifies the specifications of the applications. This can be accomplished by preparing a standard application definition form that contains

- Data input requirements
- Processing requirements
- Output products

In addition, the form should also identify the function in which it will be used when the system is operational. An example of an application definition form is shown in Figure 5.8.

Once all the applications for the system have been defined in this manner, the planning process can begin to analyze the scope of the GIS in order to structure the approach needed to schedule the remaining phases of the project. Large multiparticipant projects may define hundreds of applications during this process because of the diverse functions represented by the participating agencies. Even single-agency projects with diverse functional units participating can have a large number of applications defined after the analysis of processing needs. At this stage of the planning process—now that the data and the applications are known and documented—it is necessary to review these needs and group them into phases so that an orderly schedule can be developed for assigning resources and responsibilities, establishing milestones for managing the implementation process, and allocating funds throughout the multiyear implementation. Since all applications cannot be developed and implemented at the same time, it will be necessary to assign some priority or ranking so that those that are most important can be implemented before those that are not as important. This will assist in establishing a plan for database conversion, hardware and software acquisition, and other activities.

APPLICATION:

Display zoning map information for an area defined by the user.

FUNCTIONS USING THE APPLICATION:

Review and prepare zoning changes.

DESCRIPTION OF APPLICATION:

This application uses zoning and related parcel-based data from the database to display existing information related to zoning for a specific area that is defined by the user. The application must be available interactively at a work station when the user invokes a request and identifies the subject land parcel. The application will define a search area based upon the search distance defined and input by the user and will display all required data for the area within the specified distance from the outer boundary of the subject parcel.

DATA INPUTS:

User Defined:	Parcel identifier
	Search distance

Data Base:	Zoning boundaries
	Zoning dimensions
	Zoning codes
	Parcel boundaries
	Parcel dimensions
	Parcel numbers
	Street names
	Addresses

PRODUCTS OUTPUT:

1.   Zoning map screen display with subject parcel highlighted, search area boundaries, search distance, all zoning data, parcel data, street names, and addresses.

2.   Hard copy map of the above.

FIGURE 5.8 Sample applications definition form.

The setting of priorities or rankings can be a difficult task because of the number of different agencies and functional units involved in the project. With many different personalities, funding sources and amounts, political and organizational environments, and other related influences affecting the system design process, effective project leadership and communication is important during the setting of application priorities.

One way to avoid controversies and gain consensus during the establishment of application priorities is to obtain agreement beforehand on the criteria that should be used to measure the importance or priority of each application. In the following is a recommended grouping of priorities into three categories (Croswell, 1988):

1. *High-priority applications* are extremely important and may be defined as those that

   - Have the greatest impact upon all participants
   - Are the most often used in the day-to-day operations of the agencies
   - Are the most labor-intensive when done manually or
   - Are currently experiencing the most problems that have a direct effect on a major function

2. *Moderate-priority applications* are important, but do not have the urgency that warrants a high priority. They may be characterized by conditions that

   - Provide general capabilities that can apply to more than one type of data or situation
   - Affect standard functions that do not have a high volume of use or
   - Affect functions that do not have a major impact on the daily operations of the agencies

3. *Low-priority applications* have the lowest impact on the success of the system because they

   - Provide capabilities that affect only one or a small number of functional units
   - Provide a capability that cannot be achieved until another application is implemented
   - Provide a capability that is not needed immediately, but will be at some future time or
   - Enhance a function by creating a new capability that is not currently performed

The establishment of application priorities need not result in a rigid schedule that governs exactly when each application will be implemented. This is because the analysis, installation, and base map creation process consume a large amount of time, and conditions affecting certain applications are likely to change during the course of project development. These priorities do, however, provide guidance and direction for the establishment of hardware, software, and database requirements, which is helpful when scheduling resources and in providing estimates for costs and benefits over a multiyear time frame.

Hardware and Software Requirements

Once it is known which *functions* of the organization will be using the GIS, what the content of the *database* will be, and what *applications* are needed to support the processing of the data for these functions, it is possible to begin to specify the hardware and software needs of the system. This is because the functions define where the hardware components will be needed; the database content defines the size, source, and update volumes of the data storage components; and the applications form the basis of the software capabilities required of the system.

In all information systems development projects, the applications define the software capabilities that are needed; the software defines which types and mixture of hardware components must be installed. The GIS, while complicated by the spatial nature of the data and the breadth of involvement by many participants, is no different. There are more than 60 vendors of GIS hardware and software on the market today, ranging from single-purpose, microcomputer-based technology to large workstation networks and mainframe-based systems. It is virtually impossible to select the most appropriate system without first analyzing them in terms of the important applications for the particular GIS being planned.

Software

Generally speaking, GIS software capabilities can be grouped into three functional classifications:

- Automated mapping functions
- Data management functions
- Spatial analysis functions

Each application should be reviewed with respect to their need for these three types of functions.

Automated Mapping Functions. These functions manipulate the cartographic records of the GIS for the purpose of extracting, updating, creating, and producing high-quality maps and drawings. Their focus is on the specific mapping process itself. Thus, applications that place a high importance on mapping and drafting operations may require such functions as coordinate transformation, map scale conversion, coordinate geometry (COGO), edge matching, windowing, curve fitting, area calculation, line-length calculation, text placement, snapping, copy parallel, precision entry, and rubber sheeting.

Data Management Functions. These functions manipulate the nongraphics data stored in the databases of the GIS. They create and update data, retrieve and manipulate selected records, and produce standard and ad hoc reports. Applications with a heavy dependence on attribute data may require such functions as ad hoc inquiry, ad hoc report generation, summarization, security, Boolean logic, and standard data entry forms.

Spatial Analysis Functions. These functions utilize both the cartographic data and the attribute data for processing in a spatial context, often with topological relationships. These functions produce results of analyses in a statistical nature and often create new maps or new databases. Applications requiring spatial analysis capabilities may require such functions as proximal analysis, network analysis, polygon overlay, point-in-polygon, choroplethic mapping, buffering, and spatial aggregation.

In addition to the software functions mentioned, there are other software functions that may be required that are of a more general nature or address a unique situation or that may be required because of the current computing environment that is already in place and planned as a resource in the new system.

Communications Software. If data files must be transferred between workstations or if different software is needed by different users, then it may be advantageous to take advantage of communications software in a network environment.

Menu Design. Many sophisticated commercial systems offer a facility whereby the menus used to supply commands to the system (either on a tablet, keyboard, or screen display) can be tailored for a specific application at a specific user workstation. Programming these menus allows the user to select certain options to perform system functions so that individual commands do not have to be used.

Symbology Creation. In the design of specific map outputs, it will be necessary to establish standard symbols, text fonts, and line symbology (such as used in portraying water main valves, land use symbols, and railroad lines). While most systems provide a range of "built-in" symbols for many features, they also provide a capability to design specific symbols and fonts that may be desired for a unique situation or installation.

Interfaces with Existing Packages. If a specific database management system is currently being used and contains data for the system, or if the system must interface directly with an installed computer-assisted mass appraisal system, computer-aided dispatch system, or other special-purpose system, then these requirements must be identified prior to the selection of the software for the GIS.

Other Customized Programming. In addition to software needs that may be unique to an installation, there are other special programming needs that may be required, such as standard map designs, special macrolevel programs, and file archiving programs.

Hardware

By waiting until after the data, applications, and software needs of a GIS are determined, defining what hardware components are needed can be a fairly uncomplicated task. The easiest approach is to analyze hardware needs in a backwards fashion: determining what devices are needed by the users whose applications will be implemented in accordance with the plan. By looking at the input and output needs of the applications for each user (see Figure 5.8), the specific hardware components needed for those users can be identified:

- Workstations
- Plotters
- Digitizers
- Printers
- Scanners
- Alphanumeric terminals

A brief discussion of these input/output devices will shed some light on the various criteria that should be applied to their selection.

Workstations. There are two general options between which to choose when evaluating the need for the computer workstations used in each application: a terminal connected to a centralized computer (such as a microcomputer, minicomputer, or a mainframe) that may also have additional terminals connected to it, or an intelligent workstation that is itself a computer and that is connected to other intelligent workstations and computers on a communications network. The criteria used to determine which configuration is most appropriate are based upon software functionality, system hardware use, and data file availability.

In a centralized computer system with one or many terminals connected to a single computer, all software and data files reside on the computer and all terminals use the same software and data files. This configuration also usually requires plotters to be connected directly to the computer (although some lower-quality hardcopy output devices can be connected directly to the terminal), and the central software is used to produce the plots. The computer may also be connected to another computer so that data files can be transferred (or even update and inquiry transactions), but the speed of these communications can be limited.

The network configuration allows a much more versatile use of software, hardware, and data files. Each computer (micro, mini, or mainframe) connected to the network can operate independently of the other computers (much the same as with the centralized configuration) with its own software, hardware devices, and data files; however, each computer can also have access to the data files, software, and hardware devices of the other computers on the network. This can vastly improve system response time because the files are not transferred between computers—access is direct (as if the file actually resides on the workstation attempting access). Outputs can be directed to any plotter on the network, regardless of the computer to which it is connected, and software can be unique to each computer on the network, allowing a data file to be used by different software in different functional units.

So, the determination of the type of workstation needed by the user depends upon whether a versatile computing environment is needed. Where all users need the same software functionality, the same databases, and are located close together without being physically dispersed (so that they have easy access to a plotter and other centrally located devices), then a centralized computer system may be appropriate. This configuration is often less expensive than a network-based system and is less complex, requiring less technical expertise and support on an on-going basis. If, however, users have a variety of diverse software needs (automated mapping, database query and manipulation, spatial analysis), anticipate a wide variety of different databases, each maintained separately by different functional units, and also will be physically separated at different sites, then a network-based system may be more appropriate.

It is also important to consider future needs in deciding upon the hardware configuration needed—especially those applications that were given a low priority—because the use of the system is most likely to expand once the initial applications are implemented.

Plotters. Again, the applications planned for implementation on the system should determine the type and number of plotting devices needed in the system. Since plotters are high-cost components of a system, it is important to balance the cost associated with the various features of different plotters with the specific needs of each application. Variable features of different plotters include size, quality, and speed of output as well as color. Applications that require high-quality output production may require a pen plotter unless the users responsible for the output are satisfied with the quality of an electrostatic plotter. (A sample output from each type is usually necessary for such determination.) Consideration for the degree of output quality, however, must be balanced with the speed of output (pen plotters are slower) as well as ongoing maintenance (pen plotters require constant attention and maintenance of pens, ink, and other components). Applications that require a high volume of output should consider the use of an electrostatic plotter since pen plotters can be as much as 10 to 100 times slower. For applications with high-volume and high-quality output requirements, it may be beneficial to consider a computer output microfilm (COM) device that produces the final plot directly onto film for later processing onto paper medium. Color may be important in some applications (such as highlighting certain features for edit checking or for producing presentation-style maps for communicating the results of analyses); however, color plotters are more expensive and also require more attention and maintenance. Applications that do not require large plots or high-quality output, but do need some hardcopy product, may be able to make use of low-cost laser printers that can be connected directly to a workstation for easy access.

Digitizers. Large digitizing tables are needed only in certain situations and applications. During the initial database creation, it may be necessary to use such tables if existing map products are used to build the database. Later, after the database is completed, these large tables are not as necessary since updates can be made directly at the workstation screen with a smaller, less expensive editing digitizing tablet.

Summary

This chapter has addressed a methodology for designing a geographic information system from a functional point of view: how the functions of an organization determine the data and the processing of the data that is needed for successful use of GIS technology. It has discussed the development of a current physical model of the organization that documents all major functions of the organization to ensure that the GIS design can be comprehensive and responsive to needs regardless of the organizational structures that may be in place at the time of the analysis. By decomposing the organization and its functions into their principal components (processes), it is then possible to analyze information needs of the organization from an independent point of view—a new logical model that is an abstraction of current organizational structures based upon what the organization does and what information is needed to do it. This method ensures an appropriate match between the technology and the organizational environment in which it must operate to be used successfully.

This chapter has also presented an approach for determining the map and attribute information that is needed to support the functions of the organization and the GIS applications, hardware, and software that must be implemented to be able to process the information.

We have now explored the design aspects of the GIS project that are needed in order to prepare specifications in selecting a specific system. When it is time to implement the design, additional needs (staffing, training, procedures, and organizational, institutional, and legal needs) must be addressed, as identified during the implementation planning process.

NOTES

1. This old French saying, translated as "You shouldn't sell the bearskin before killing the bear" underscores the need to time the various steps in any project properly. It applies well to the planning process for development of a geographic information system and the procurement of hardware, software, and data conversion services. It seems like common sense to suggest that an organization should know what they are going to do with a system before proceeding with GIS implementation, but in reality, this has not been the case with many groups that have gone ahead with detailed and expensive procurement processes. In many cases, the predominant philosophy has been one of "let's get the hardware and software, start loading data and decide what we are going to do with the system later."

2. Coordinate Geometry (COGO) is a program used to generate digital map features from geometric descriptions. Mathematical algorithms compute coordinates from geometric descriptions such as bearings and distances; the coordinates are stored and used to generate graphic map displays. COGO has become the common shorthand name for the technique (Antenucci, 1991).

3. Such can be the case, for example, in the simple process of geocoding address-based records. This application need not require digital property maps that can take years to create—it needs only an adequate geographic base file.

4. Public works departments, for example, are normally subdivided into smaller bureaus or divisions such as sanitation, forestry, and street maintenance. Likewise, a public utility may have a large construction division subdivided into responsibilities for different facilities: distribution systems, plant facilities and so on.

REFERENCES

Antenucci, John (1990). GIS implementation: Getting started. In *Government Technology*. GT Publications, Sacramento, Calif., pp. 20–22.

Antenucci, J., K. Brown, P. Croswell, and J. Kevany, with H. Archer (1991). *Geographic Information Systems: A Guide to the Technology*. Van Nostrand Reinhold, New York.

Arthur Young & Company (1987). *The Arthur Young Practical Guide to Information Engineering*. Wiley, New York.

Brown, P., and D. Friedley (1988). Assessing organizational preparedness for a comprehensive, distributed LIS/GIS. In *Papers from the 1988 Annual Conference of the Urban and Regional Information Systems Association,* vol. 3, pp. 22–36.

Campbell, H. (1991). Organisational issues and the utilisation of geographic information systems. In *Regional Research Laboratory Initiative Discussion Paper Number 9,* Department of Town and Regional Planning, University of Sheffield, Sheffield, Eng.

Croswell, P. (1988). Definition of applications as a basis for GIS planning and system procurement. In *Papers from the 1988 Annual Conference of the Urban and Regional Systems Association,* vol. 2, p. 13.

Finch, R., H. Jones, and J. Litterer (1976). *Managing for Organizational Effectiveness: An Experimental Approach*. McGraw-Hill, New York.

Huxhold, William E. (1991). *An Introduction to Urban Geographic Information Systems*. Oxford University Press, New York.

Huxhold, William E. (1993). The application of research and development from the information systems field to GIS implementation in local government: Some theories on successful adoption and use of GIS technology. In *Diffusion and Use of Geographic Information Technologies,* Ian Masser and Harlan J. Onsrud (eds.), chap. 1.4. Kluwer, Deventer.

International Association of Assessing Officers (1989). GIS Guidelines Questions, internal working document, September.

Martin, J., and C. McClure (1983). *Software Maintenance.* Prentice-Hall, Englewood Cliffs, N.J.

National Research Council Panel on a Multipurpose Cadastre, Committee on Geodesy, Assembly of Mathematical and Physical Sciences (1980). *Need for a Multipurpose Cadastre.* National Academy Press, Washington, D.C.

ADDITIONAL READINGS

Croswell, P. (1991). Obstacles to GIS implementation and guidelines to increase the opportunities for success. *Journal of the Urban and Regional Information Systems Association* 3, no. 1:43–56.

Dangermond, J. (1990). The organizational impact of GIS Technology. *ARC News,* Summer.

Evans, D. (1987). Distributed data processing in a municipality: How to make it a win–win proposition. In *Papers from the 1987 Annual Conference of the Urban and Regional Information Systems Association,* vol. 3, pp. 157–68.

Ezigbalike, I., D. Coleman, R. Cooper, and J. McLaughlin (1988). A land information system development methodology. In *Papers from the 1988 Annual Conference of the Urban and Regional Information Systems Association,* vol. 3, pp. 282–96.

Gibson, C., C. Singer, A. Schindman, and T. Davenport (1984). Strategies for making an information system fit your organization. *Management Review* 73, no. 1:8–14.

Huxhold, William E. (1992). Needs assessment. In *Multipurpose Land Information Systems: The Guidebook,* D. David Moyer (ed.), chap. 16. National Geodetic Survey, North American Atmospheric and Oceanic Administration, Washington, D.C.

6

Implementation Management

The best managers think of themselves as playing coaches. They should be the first on the field in the morning and the last to leave it at night. No job is too menial for him if it helps one of his players advance toward his objective. How many times has a critical project been held up because there was no one around who could get someone out of bed, or type up a fresh draft, or run off some copies on the Xerox?

ROBERT TOWNSEND

The previous three chapters have focused on strategic and implementation planning, conducting a needs analysis, and defining a design methodology. Our focus now changes to applying that knowledge to the creation of a working system. A colleague of ours once pointed out that "a pile of design documents, no matter how complete, is not a working system." It has been argued that a well-established understanding of the organization and its needs with respect to GIS are essential for successful GIS implementation. While this remains true, it is this next stage of work that represents the majority of effort and the one in which most of the decisions will be made.[1]

Following the outlined process (as presented in Figure 6.1),[2] the project will have progressed to a stage between being generally designed and being operational. Depending on the scope of the implementation and the approach being followed, the GIS may exist in this phase of development for some time, perhaps years. Various parts of the GIS may be in operation while others are still being designed (if an incremental design approach is being followed).

This is a period of transition, from planning to doing, from abstract to concrete, during which concepts and plans are translated into working systems. Implementation decisions heretofore will result in data being collected and converted, staff being hired or trained and reassigned, changes being made to business procedures and perhaps to organization structure, technology purchases, and applications construction.

Management emphasis will also change from planning to the creation of products ("deliverables"). A new or revised management framework will be

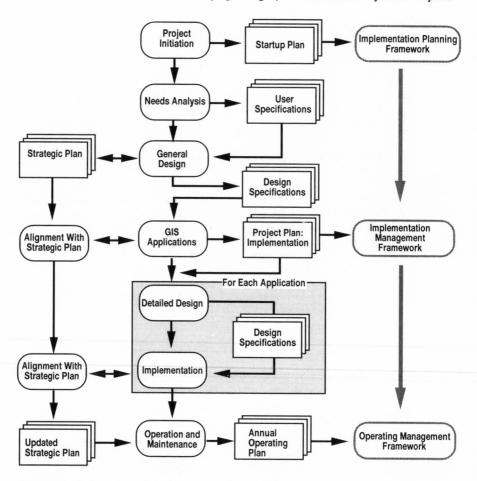

FIGURE 6.1 Review of the design and implementation process.

instituted. The user groups who have actively ushered the process through planning and analysis will now hand over most tasks to technical staff. Changes to procedures and organization structure that were agreed to during planning are now becoming reality.

There is at this point a danger that some of the commitment and momentum generated in previous phases will wane. There may be a transfer of responsibility and possibly a change in project management. This is a good point to stop and regroup. End users, management, and project staff should meet and review what has been accomplished and what remains to be done. The strategic plan should be reviewed and revised if necessary. Expectations and resource allocations should be reviewed and realigned, if necessary, as part of the process of reviewing the strategic plan. Unless there have been major new discoveries during the needs analysis, then this should be a relatively easy task. New participants in the process (new technical staff, contractors, and possibly technology vendors) should be introduced and acquainted with the status of the project.

The needs analysis and general design studies will have defined

- The applications to be implemented and their priority and sequence
- Major organizational and practical constraints to implementation
- The data and processing requirements for each application (functional specifications)
- A general design specification for the system including the general hardware and software requirements to support each application and an overall technology architecture
- A design methodology to direct detailed design and implementation

These will be described in the general design specifications and the implementation plan. The implementation plan will also have defined the requirements for the management framework for directing and administering implementation, but it will perhaps not be instituted yet.

Detailed design and implementation will begin during a period of transition from planning to doing. Much more is now known about the requirements, opportunities, and costs. Some assumptions and concepts have been confirmed, while others will have been refuted. For many GIS projects, a major decision point will have been reached. Management must now confirm and finally commit to

- The applications to be developed
- The resources required for implementation
- The sequence of development
- Who will perform the work (internal staff, contract staff, or some combination)
- The selection and assignment of staff
- A host of related decisions, including which to delegate and to whom

Management will also need to deal with the impacts from these decisions. Even with adequate participation and successful management of the planning process, some participants will be disappointed with the decisions made. Additionally, effects of change that were distant and abstract during planning are now becoming a reality. Management will need to be sensitive to the various people issues that will be encountered during implementation to ensure continued staff support.

Following rapidly behind these people and organizational issues will be another extensive series of technical and project management decisions. Once again, it is vital that an appropriate management framework be established to deal with these issues.

These and related topics are the subject of this chapter, beginning with management decisions and implications.

Managing the Transition to an Operational System

The transition from planning to construction marks a major milestone in the implementation process. There are many decisions that must be made at this point, including: conformance with plan; managing change; establishing a new project management framework and managing the transition; hiring and reassigning staff and the associated training; procurement management; and contracting practices.

The introduction of information technologies of any sort brings about change in the organization into which it is introduced.[3] This is a well-supported fact that will be truly independent of the specific choices about the approach to implementation.

Change and the importance of process in dealing with change was stressed in earlier chapters, but from a process design and planning perspective. The impacts of the prescribed changes will materialize during implementation. It is natural that participants will experience some uncertainty and unease at this point. How this is dealt with will determine the impacts and subsequent future support for the GIS and associated changes to structure, procedures, and job responsibilities.

There are two forms of transition that will occur as the GIS moves toward becoming operational: changes to the project management framework and organizational change to facilitate implementation, and organizational change resulting from implementation of the GIS.

The Management Framework developed for planning will need to change somewhat and additional staff will need to be hired or reassigned, at least for the period during which implementation occurs. It is normal for the organization to be in a period of transition at this point. The manner in which these changes are dealt with will determine whether the commitment and enthusiasm fostered during earlier work is maintained through implementation.

Organizational changes will result from and in support of the GIS implementation. Fundamental to the GIS is the shared database (if not physically, then certainly conceptually).[4] In many organizations this will present a new data management paradigm with its associated changes in procedures, work groups, and individual roles and responsibilities. Persons previously responsible for isolated well-defined tasks, will now need to broaden their outlook, recognizing their contribution as part of a larger, likely networked process. Likewise, management will need to recognize these changes through training, career planning, and compensation. For many staff the period of transition from the pre-GIS environment to the post-GIS environment will mean additional work, stress, and commitment. Such added contribution by staff must also be recognized. This is particularly true if an incremental implementation approach is used. If staff participating in early applications are not treated well, it will be very difficult to elicit enthusiasm and commitment from staff participating in later applications.

Issues that were straightforward while conceptual and distant become real and immediate during implementation. Staff changes, procedural changes, and so forth may become sources of opposition. It is important that these are dealt with openly and with adequate participation by staff and their managers who will be affected by the changes. Hersey and Blanchard (1982, pp. 273–74) define two types of change process: participative and directive. The stages in each process are illustrated in Figure 6.2.

Participative change occurs when new knowledge is accepted by an individual or group and a commitment is made in the direction of the desired change. Participative change requires individuals and groups to be achievement motivated and for the organization to have an environment that encourages action by the individual. The power to facilitate the change largely resides with individuals alone or as part of a group.

Directive change in its most direct form consists of commands and edicts. It is the result of power from position: the "boss." Management declares that a change will occur and expects staff to comply. In fact staff usually do "comply," which, as discussed earlier, is the problem. The organization needs commitment, not compliance, for GIS implementation to be successful.[5] More positively, directive change can consist of leadership and guidance to initiate participative self-directed change. An appropriate amount and format of directive change fosters group coordination and commitment accelerating participative change. To initiate the process, leadership, and perhaps direct involvement, by senior management, may be required to foster the appropriate environment. Successful change is most likely to result from an appropriate balance of directive and participative change.

A discussed in Chapter 4, participation, appropriate leadership, and support, will determine whether changes are met with compliance or commitment. Compliance with change will mean low levels of support, lack of innovation, and reduced or nonsustainable benefits. To a large extent the difference between a compliant and a committed behavior rests with the working environment and

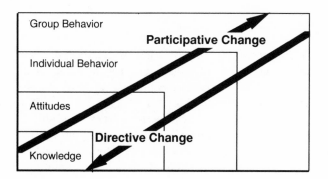

FIGURE 6.2 Participative and directive change.

support from the organization. Creating supportive working environments that are not threatening will reduce compliant and resistant behavior.[6]

Dickson and Simmons (1970) proposed seven procedures to reduce behavioral problems associated with the introduction of information systems:

- A positive organizational climate with feelings of trust, cooperation, communication and full management support for the system
- Participation in decisions pertaining to the system by all employees who will be affected
- Clearly stated characteristics and purpose of the system
- A willingness to weigh individual needs against system efficiency
- Emphasis on new, challenging job tasks to replace those taken over by the computer
- Establishment of new criteria for performance evaluation
- Tailoring of output to meet users' needs

Having involved the end users of the GIS and their management from the beginning is perhaps the most effective and least costly way of achieving the necessary commitment and support for change. However, there are additional opportunities to deal with these issues during the creation of the implementation management framework, when new staff are hired, and during training.

Organizational change will continue throughout implementation until the system becomes stable. Organization issues have been dealt with in all previous chapters and they will emerge again in Chapter 7. Different stages have different needs as the emerging GIS organization grows in size and permanence.

In larger organizations there may be a need to involve the human resources department to obtain support for training, career development, new position descriptions, and special considerations for compensation and benefits. If this is the case it is advisable to have the human resources department actively participating in the management framework. This subject will be pursued further during the discussion of staff and training. However, before we do that, let us look at some other decisions requiring management involvement.

Outsourcing GIS Implementation

Having developed a clear picture of the needs, issues, costs, benefits, and capabilities of the organization, one of the decisions that management must make is determining the optimum mix of internal and contracted activity. At one extreme, all work is performed within the organization by organization staff; at the other extreme, all GIS implementation and perhaps operation will be contracted to an outside services firm—*outsourcing*. Outsourcing, also known as facilities management, has in recent years become an accepted means for organizations to deal with their information technology needs. Outsourcing is a contractual arrangement in which an external company takes over the information systems operation, delivery of services, and sometimes planning of an organization.[7] There is considerable debate now about the benefits of out-sourcing. We will not get into that debate. Interested readers are referred to sources in the notes.

While not common in the GIS arena, there are a few examples. The government of Ontario, Canada, has entered into a joint venture with a consortium of private companies for the purposes of constructing and distribut-ing digital cadastral data for the Province.[8] It is certain with companies like EDS that are active providers of outsourcing services getting into GIS (they purchased the McDonnell Douglas GDS system product) that we might expect to see more outsourcing associated with GIS.

It is more likely that some rather than all the work will be contracted to outside consultants and contractors. The following are some examples of when contracting may be used to support the GIS implementation:

- When the skills and expertise required to perform a particular task are not available from existing staff but will not be required after completion of GIS implementation
- When current work loads prevent internal staff from performing GIS tasks: Contractors are retained to perform current work while regular staff participate in GIS implementation
- When the cost of completing a task is significantly lower if the task is contracted out—for example, data conversion can frequently be per-formed at lower costs by firms specializing in data conversion because of specially tailored methodologies, and because frequently much of the data conversion work is performed at the firm's offices in countries with much lower labor charges
- When specialized technical expertise is only available from contractors specializing in that work
- When products, or the customization of products, can be purchased more cost effectively than they can be developed internally: Many technology vendors (system integrators) will assemble technology to custom specifications at costs that are lower than could be achieved using internal staff (when all factors such as technical risk and downtime are consid-ered); additionally, these same firms may contract to maintain and upgrade the systems (this could be considered as a form of outsourcing of the information systems facilities)

Most organizations will undertake implementation using a combination of internal staff and contracted services. However, the mix will vary by organiza-tion. Consequently, for the balance of this discussion, it will be assumed that most activities are performed by the organization so that all subjects can be covered.

Staff and Training Needs

Building and operating the GIS will require full-time staff with specialized skills. New staff may be hired or contracted, and existing management and staff will require education and training. The related activities must be scheduled and performed in concert with technical and other management activities.

The GIS organization—staff, skills, policies, practices—will be built gradually. The GIS organization will evolve as new staff and new skills are added. The longer-term organizational needs for GIS administration will depend, in part, on decisions made about the number and type of applications, and on how much of the development will be contracted to external services, as discussed in the previous section. It is not necessary that all the staff and skills outlined in this section are available immediately. It takes considerable time to "grow" a good GIS organization. The GIS organization will continue to grow and develop as applications become operational. (The subject of GIS staff will be dealt with further in Chapter 7.)

While GIS needs and applications will be specific to each organization, some general requirements are common to all GIS implementation. On that basis, it is possible to make some general statements about staff.

Most GIS applications will include the capability to determine the location of a land-based activity, to determine characteristics of the activity and the location, and possibly to assess the impact of the proposed activity on the existing land characteristics. Within this context, the operations and functions that must be supported by an GIS are

- Processes for collection, compilation, and formatting of data for storage
- Mechanisms for secure, orderly storage of the data
- A maintenance process to keep the data current
- Techniques and procedures for analyzing the data to extract the information required for management decision-making
- Procedures and standards for presentation of the information in the format required for use in decision making
- Mechanisms for delivery and use of the information products

The preceding functions are sufficiently general to describe most information-handling processes. The management of land and resource information is further complicated by the inclusion of a geographic component, and the resultant requirement for computerized mapping (the need for map output is assumed). It is these specialized processes that create the need for specialized staff and the associated organizational structure to optimize the effectiveness of the GIS.

Somewhat independent of the organizational approach adopted, certain skills and staff positions are required for GIS implementation and operation. The required roles and expertise are

- Project sponsors (senior management)

- Business (line) management
- Project management
- GIS applications design and programming expertise
- Geographic data management and display expertise
- Technology implementation and operation expertise
- End users

Note that in the foregoing list both management and end users have been included as "staff" in the GIS implementation. While end users and management may not be directly involved in the construction, they are important to the administration and management of the construction process.

The exact skills, and number of staff with such skills, will depend on the size of the system (number of applications), contracting decisions (how much of the work is performed internal to the organization), the technology used (hardware and software architecture), and the scale of the operation (the size and diversity of the database and the number of users).

Senior Management: Project Sponsors

Senior management must develop enough of an understanding of the GIS project to assess whether it is progressing as intended: not only from the standpoint of monitoring progress, but also to ensure that there is continued commitment to the project as conceived. Projects frequently lose senior management commitment because there was an inadequate understanding at the outset, of the time lines and of the costs for implementation. Management develops the perception that the project has gone wrong, when actually it is proceeding as planned, just not as understood by senior management. Senior management must therefore take the time to develop an adequate understanding of project objectives, plans, and budgets to assess project progress effectively. Additionally, any major changes in direction involving outsourcing or other equally significant policy shift can only be made by senior management.

GIS Management

Any major project needs a full-time project manager. It seems almost trite to state this, but experience has shown that many projects attempt to function with part-time managers. In this way the responsibility for GIS implementation is added to a line manager's work. Line management should not be confused with project management; line management is business management. It involves all aspects of the business including personnel, finances, and ensuring profitability or the mandated equivalent in a government organization. The line manager's job existed before the GIS and presumably will continue after implementation (unless the functions are being replaced by the GIS, which is uncommon). Therefore the line manager will deal with the GIS project as one among many for which he or she is responsible. Successful GIS implementation requires dedicated attention to a myriad of details. Additionally, special project management and systems analysis skills are required for success. The issue of GIS management must be resolved for a GIS implementation to be successful.

During the implementation planning process someone has likely emerged as being responsible for the "GIS project." The responsibility may have been shared by several managers composing a committee, but it is likely that someone has led GIS development (a GIS champion perhaps). However, it is also likely that GIS management was only one of several of the manager's responsibility areas. If a suitable project manager was hired at the outset of the planning phase, then project management may not be as much of an issue now.

However, if a permanent manager has not been assigned, now is the time to make some deliberate decisions about GIS management. The responsibilities for GIS management need to be formalized and translated into a full-time position or positions (depending on the size and complexity of the project). From that point forward there will be an ongoing need to have management-level staff to plan, budget, and schedule GIS-related activities, and to manage technical and other details associated with GIS implementation.

During implementation, there are two key responsibility areas that must be satisfied: *GIS administrative management* and *GIS project management*. GIS administrative management entails budget planning, human resources management, alignment of GIS activities with other business activities of the organization, and perhaps broader information systems planning and management, depending on the organizational placement of the manager and the scale of the project (see discussion in Chapter 7). This person will be called the *GIS business manager*.

GIS business managers should possess a conceptual understanding of GIS principles and technology, have participated in (if not led) the planning process, have knowledge of the business functions to be supported by the GIS (e.g., environmental assessment, building permits, soil survey, forest inventory), and have experience with administrative management, information systems development, and personnel management. Perhaps most important is a firm understanding of the organization's culture, its goals, and its operating policies. The principal responsibility of the GIS business manager will be to ensure that the GIS is appropriately applied to the organization's business functions and that the GIS project has the necessary support from the organization to succeed. Business managers should be staff positions drawing from existing management, or possibly promoting qualified staff from within the organization. All organizations may not have a suitable individual, and it may be necessary to develop someone, "growing" that person along with GIS implementation. In large projects, GIS business management will require a full-time position. In smaller projects, GIS business management may be performed under an existing line management function.

There will be a separate need for a *GIS project manager* and or a *project leader* (or on very large projects, a project management team). During construction of applications and implementation of other components of the GIS, the prime function of the project manager is to deliver a successful system on time and within budget. The skills required are: project planning and scheduling, cost control, personnel management, GIS and other technology design experience, and computer systems implementation experience. Unlike the business manager, this position can be contracted by selecting specifically qualified individuals to deal with a specific project or with a project component.

The project manager should *ideally* possess a broad range of skills including: some understanding of surveying and mapping, thematic data collection (e.g., soil survey, vegetation inventory), administrative management, contract law, information systems planning and design, design methodologies and tools, personnel management, and modeling with geographic data. The best personal attributes include diplomacy, good written and verbal communication, and a relentless drive for success. The project manager must understand the work to be done and be able to assign pieces of the work to the individuals most capable of performing each piece. The project manager must arrange for the project staff to have the environment and tools they need to perform required work.

Finally, the project manager must routinely perform what Metzger (1981) calls "sanity maintenance"—to step back from the details of the project to evaluate status. Sanity maintenance includes learning to say "no" to divergent requests. It involves ignoring some issues, because they will literally go away. Most of all, sanity maintenance means assigning priorities to ensure project success, keeping project staff happy and project progress at the top of the list.

There is a difference between the roles of business managers, project managers, and project leaders, as demonstrated in Table 6.1. GIS business management is not the same as project management. Business management involves all aspects of the business, including personnel, finances, and ensuring that the objectives of the organization are met. The project manager's only focus is the successful implementation of the system. Special project management and systems analysis skills are usually required for success. Project management can be frustrating, and the position requires considerable skills in negotiation, interpersonal communication, and past experience in the design and implementation of automated information systems. The project manager must plan what is to happen, obtain agreement, and ensure that what has been planned happens. For these reasons, project management is often contracted out.

Project Leader

In situations where there is considerable custom development to be performed and where the work is to be performed within the organization, there will be a need for a project leader in addition to management staff. The project leader is a senior technical person of management caliber who is primarily oriented towards technical issues and supervision of technical systems staff (programmers). In large development projects this person is sometimes referred to as the *chief programmer*.

Table 6.1   Comparison of Business and Project Managment

	BUSINESS MANAGEMENT	PROJECT MANAGEMENT
Type of management	Business Personnel	Technical
Resources	Resource control	Negotiates for resources
Level of authority	Direct	As required for project
	Delegates responsibility	Advises line management
Type of position	Permanent	Temporary
Focus	Project just one task of many	Project only task, singular purpose

Although the focus of the project manager and project leader are different, their goal is the same: successful implementation of the GIS. The bond between project manager and project leader is made up of shared trust and respect. Each must believe the other is doing the best he or she can to make the project a success. Each must believe the other is truly concerned with the staff doing the work. Each must be ready to accept criticism and suggestions from the other. The project manager and project leader must work in tandem. As shown in Figure 6.3, some roles and responsibilities are shared (as represented by those placed in the middle of the spectrum), while the dominant role for others lie more with either the project manager or the project leader (as represented by those on either end of the spectrum.

Other GIS Staff

At various points during GIS implementation (again depending on the amount of work performed internal to the organization and the size of the project), the organization will have need for the following other staff and skills:

- One or more system/data administrator(s)
- GIS analysts
- Programmers—database and GIS
- Production personnel, including cartographers, technicians, and data entry clerks and supervisors

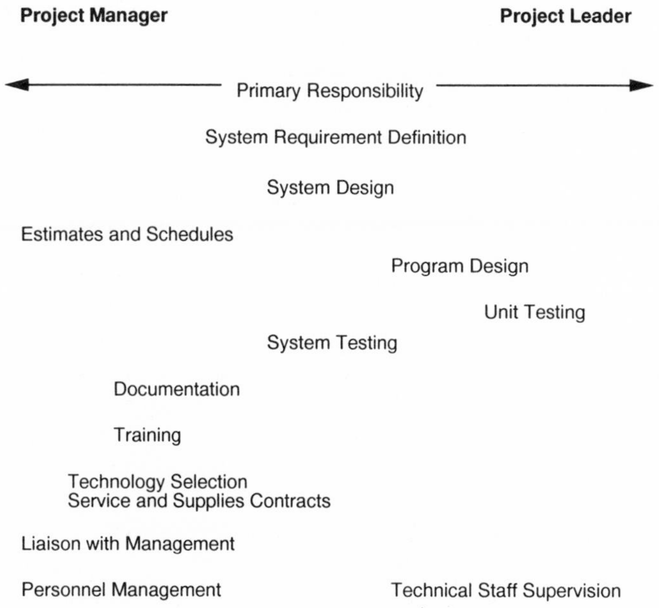

Project Manager   **Project Leader**

◄──────────────────── Primary Responsibility ────────────────────►

System Requirement Definition

System Design

Estimates and Schedules

Program Design

Unit Testing

System Testing

Documentation

Training

Technology Selection
Service and Supplies Contracts

Liaison with Management

Personnel Management   Technical Staff Supervision

FIGURE 6.3 Comparison of roles and responsibilities of project managers and project leaders.

- Support staff—administrative and secretarial
- Users/coordinators

There are many ways the organization can obtain the required skills. As discussed earlier, they need not all be in-house staff. Also, the number, type, and sophistication of skills of staff will vary with the size of project and will grow in number as the system expands. Other factors that influence the number of staff are current work loads, how geographically distributed the system is (different offices), and the skills of individual staff members. In addition, more experienced staff are required for development and implementation than for ongoing maintenance and support. Finally, not all the positions described here are required to get started.

Appropriately skilled staff can be hired, trained, or contracted. Which option is best depends on the position, the availability of skilled staff, and for how long the skills are required. Suggested approaches are given for each of the positions in the following.

Computer System/Database Administrator. This is the senior technical position responsible for technical management of the GIS, including

- Administration of the database, including maintenance of the database design, data integrity, security, and disaster recovery

- Management of routine aspects of the computer systems, such as ensuring that regular backups are performed
- Software installation and maintenance
- Design and creation of security and user accounts
- Coordinate hardware maintenance and discussions with hardware vendor representatives
- Fix or determine the means for fixing technical problems that will arise

The knowledge required is an understanding of software system directories and the organization and use of the computer operating system, the GIS software, and other installed peripheral software. Knowledge of on-site hardware, including graphics devices and digitizing tablets, is also needed.

For smaller systems, or in the early stages of implementation of a larger system, one person may be able to perform all the tasks. In larger systems there will be a need to separate the system administration and data administration tasks. In more complex implementations, the data administration task will also require specialized skills possessed only by someone who has studied to be a database administrator (DBA). Some organizations will already have a corporate database administrator or data administration group. In that event, the GIS data administration tasks can be added, or a DBA specializing in spatial data can be added to the systems staff.

Some GIS implementations will involve several sites operating over a network. Each site that has a system housing data and applications must have a system administrator. This position can be filled by an existing staff member, such as a resource professional or a senior technical person with a predisposition for computing. Past experience has shown that an interested person in this role will perform beyond any reasonable expectation for the "sheer fun of it." However, given the large amount of knowledge required for this position it is not practical to train marginal staff, and it may be necessary to hire someone for this position. During project startup, an experienced person can be contracted to set up the system and work with in-house staff, as another option.

Programmers. Programmers will be required to implement the defined applications. Two different skill sets will be required: database analyst/programmers and applications programmers. These will be the people who actually implement the GIS applications on the selected technology. While it is mandatory that these skills be available, it is not mandatory to have them as permanent staff. In many circumstances it is better that all programming be contracted for the first few years. Contracting for programming services will help to keep development on track, and to reduce the risk of development becoming side tracked by interesting diversions. It is often possible to have the vendor of the GIS software provide the necessary expertise in this area or the required skills can be contracted from another source. It may be difficult to hire appropriately trained staff and they may not be required once implementation is finished. It is not practical to attempt to train existing staff who have no or limited prior experience.

Implementation can also be structured such that on-site programming is kept to a minimum, and on-line support may be available from some vendors. If, as we advocate, applications are designed and specified in advance of purchasing a system, all of the system customization can be purchased with the system. This often results in a better system, at a lower cost. The majority of the work is performed at the vendor's offices, thus lowering the overall cost. Additionally, because the programming is "bundled" in with the system hardware and software, it is often possible to negotiate a better rate for the programming services. Performance guarantees and maintenance agreements can also be negotiated, thereby reducing risk and downstream costs.

Required expertise for the database programmer is relational database design/theory,[9] previous experience with geographic databases, and specific programming expertise in the selected database management software. GIS applications programmers must have specific knowledge of the GIS and other applications software and of how to combine various commands into finished applications that can readily be used by end users. GIS programmers should also have extensive experience with the selected computer operating system, including the operating system's user interface, graphics drivers, and related programs.

The principal responsibility of programmers is writing and testing applications software, isolating and reporting vendor software problems, and solving other software problems. Other desirable knowledge includes knowledge of map projections, scale, and other aspects of cartography as required for processing cartographic files for map production. Other duties may include managing project directories, performing complex processing functions such as spatial analysis, and setting up files for plot creation.

GIS Analyst. A GIS analyst is a staff member who assists nontechnical users with the GIS. For successful operation of a GIS it is essential to have someone with both specific knowledge of the GIS software and how to customize the GIS to perform ad hoc tasks not included in predefined applications. This position becomes more important as time passes, and more of the uses of the system become routinely integrated with the daily business of the organization (see the discussion in Chapter 7).

Specific skills that must be possessed are: specific knowledge of the selected GIS software and how it can be used; GIS concepts; database design principles; application design principles; the ability to communicate and plan specific work tasks for database creation and management, analysis, and map and report production; and a good knowledge of the computer operating system (including the programming language of the selected GIS).

Tasks that would be performed include: aiding users in identifying and defining GIS products to be generated for various purposes and translating those into processing steps for implementing the GIS application; conducting in-house software training; organizing the database; and designing and producing required maps. Other tasks that can be given to more junior staff, as available, include: the drafting and digitizing process, data preparation and map production, and photograph and map interpretation. The principal training for this work would be provided by the GIS software vendor and could be augmented with some general GIS concepts courses.

It is quite reasonable to expect to train an existing staff member who possesses the aptitude and interest for GIS work. Staff that are most successfully trained generally have a natural science, geography, or engineering background. A more experienced technician is required for development and implementation than for ongoing maintenance and support. Therefore, it may be most practical to plan to have an in-house staff member work with contracted staff during the development, than to have the in-house staff maintain what has been developed.

Cartographic Technician. This position will be responsible for a wide variety of tasks, such as map interpretation, map production, drafting, digitizing, and mapping requirements for the applications. The skills required for the cartographic technician are similar to those of the GIS analyst described, minus some of the detailed understanding of GIS and experience with defining user's needs. If a GIS analyst has been hired and the mapping requirements are not great, the GIS analyst can serve both jobs.

The system administrator can also serve as the cartographic technician, if taught to operate the peripherals and to produce the required map products. This is especially practical during startup when less production work is being performed. An additional advantage to this approach is that the system administrator can subsequently train and support future technicians. However, it should be kept in mind that cartography, computer-assisted or manual, is an art and a science in its own right. There is no substitute for a cartographic professional, if the budget can accommodate one.

The required knowledge includes: drafting techniques, the digitizing process, data and mapping requirements for a particular set of applications, photograph and map interpretation, use of peripherals (plotter, digitizer, laser printer), use of the mapping software, coordinate geometry, and map composition. The sophistication of the skills required very much depends on the business activities of the organization. The position can be filled by an appropriately skilled existing staff member (e.g., manual cartographer) trained to perform the required functions. This is particularly valid if the system administrator is experienced and can provide some support.

Clerical and Administrative Support Staff. Staff will be required to operate the system for routine functions such as database queries, entering new attribute data, and updating the database. As data conversion proceeds, clerical staff can be trained to perform table digitizing of maps and sketches. These positions can be filled by existing clerical staff trained to perform these functions. Training for these positions can occur in house as more senior staff become familiar with the technologies.

Other System Users. Not all staff who use the system will want or need to become "computer jockeys." Some staff will not be interested or predisposed to learning such skills, and the organization will not be able to train everyone, at least not in the short run. This does not exempt other users of the data products from responsibility in making the process operate efficiently. Some of the responsibilities of the other system users are

- Develop adequate knowledge of the system capabilities to plan their information needs such that the system can best respond to them
- Assist in identifying new potential applications that would be of benefit to their work and to define products to be generated
- Coordinate with the system administrators for their information product requirements
- Coordinate with other users on their shared database and software needs
- Provide support to the system staff, particularly during the early stages where staff are still learning to use the systems effectively

Economizing on Smaller Projects

To economize, it is possible to hire one person to perform several functions. For example, for smaller systems, a cartographic technician can also perform routine system administration tasks and some GIS programming. A GIS analyst can perform some system administration tasks in addition to supporting users, depending on technical expertise and workload. However, it is essential that the system administration tasks are properly and routinely performed; therefore, although it may be possible to share the work and save on staff, system administration must be a top priority for the staff responsible—the entire investment in the system depends on it.

For some positions, such as the programmer and the cartographic technician, it is useful to have highly experienced help immediately upon installing the system. There are several options for obtaining good technical support:

1. An experienced person can be hired at the outset; however, such a person may scarce and may be expensive (even so, it may be the best alternative for a longer-term position such as the cartographic technician).
2. Experienced staff can be contracted from a consulting firm to undertake the front-end development and to provide in-house training of permanent staff.
3. A contract programmer can work on site for a year or so as part of the purchase agreement with the GIS vendor.

Training and Education

Education and training have been in progress throughout the activities leading up to detailed design and implementation. The emphasis in earlier phases of work has been on building awareness through workshops and seminars to facilitate the needs analysis and general design activities. Now, however, training must be formalized. On larger projects a separate training plan (developed as part of or in context with the implementation plan) is recommended.

There are many factors to consider in designing an education and training program:

- *Who*. Senior management, business unit managers, end users, technical staff, operators, administrative staff
- *Subjects/topics*. GIS concepts, capability and limitations, use of the selected technology, fundamentals (computer cartography, geodesy, geoprocessing and spatial modeling), data management, systems administration
- *Sources of education and training*. Vendors, training consultants, campus training, on the job
- *Forums*. Short seminars, retreats, courses, workshops, conferences, on the job, self-study
- *When*. At what point in the process, at what intervals, and for how long

Table 6.2 presents who should receive education and training, the purpose of such training, what needs to be learned, when it should ideally be conducted, and recommended forums.

The top-level education is targeted for senior management and addresses the following:

- Features and benefits of a GIS and how the project will be implemented
- Impact on policies and procedures
- Impact on personnel and budgets
- Importance of senior management dedication and commitment to the project
- Where and when during GIS implementation their specific involvement is needed

The emphasis is on familiarization, impacts, and management action.

Middle management is provided with much of the same as senior management plus more specific knowledge on

- The underlying concepts and the important success factors
- Functional components of the system and how the new system will work,

Table 6.2   Sample Education and Training Program

Who	Topics	Purpose	Forum	When
Senior management process	GIS orientation Implementation	Benefits and implications of GIS implementation	Half-day seminar (GIS demonstration)	At start of GIS planning and implementation
Business unit managers	GIS orientation Implementation process GIS fundamentals/application	Familiarization Allocation of resources	In-house seminars GIS conference attendance In-house seminar	At start of GIS planning Prior to GIS implementation During Implementation
Non-technical end users training	GIS orientation Implementation process Applications limitations	Familiarization Use of GIS applications Capabilities	GIS concepts course In-house seminar Vendor training During installation	At start of GIS planning Prior to Needs Analysis
Operations staff	GIS orientation Task and technology specific training	Task of technology competence for GIS operation	GIS concepts course Vendor training On-the-job-training installation	Prior to Needs Analysis/ Functional Specs Technology
Systems staff	GIS orientation Analysis and design GIS development tools	GIS design techniques competence Software customization	GIS concepts course Vendor training Systems course	Prior to analysis and design tasks Prior to system installation and testing
Project team	GIS orientation GIS project design and management	GIS concepts GIS management competence	GIS courses Mentoring program facilitated by a GIS expert	Prior to project start Prior to GIS implementation

what is required for successful implementation, the tasks that have to be performed, and the impact on procedures and practices
- Software, for which some software-specific hands-on training may be offered as well, depending on the individuals

Operators and operations supervisors are given the most specific and detailed training.

Two levels of system use are usually offered: data entry, maintenance, and report production; and one for occasional users working through predefined menus and queries. Additionally, operations supervisors are trained in system administration.

Adequate education and training of participants will be key to successful implementation of a GIS. The budgets for training should not be an afterthought, nor should they be left to the discretion of individual user departments.

Technical training on the use and maintenance of the system is usually provided by the GIS software vendor. If possible, such training should be taken in stages to permit practice and familiarization prior to proceeding to more complex uses. This is particularly significant for the cartographic technician, programmer, and system administrator. However, training and education needs go well beyond system-specific training. GIS staff should also receive education in cartography, GIS concepts, and data management. Possessing such knowledge will facilitate changes in work procedures necessary to best take advantage of new capabilities.

Systems can be implemented so that most of the use of the system is through graphical user interfaces, with most functions available through pull-down menus and icons. This will reduce the amount of end-user training required, but may have an increased cost for custom development. There is normally a high requirement for end-user education, since many of the issues that cause discord are the result of a lack of appreciation of the complexity or magnitude of an issue. One focus of the design workshops and seminars should therefore be education. Much of the knowledge possessed by users may be related to specific GIS products other than those selected for the project. Providing users with a broader perspective will help to avoid problems stemming from product bias. There is a need to create an appreciation of conceptual and management issues for successful GIS implementation to proceed.

The basis for progressing with change is to understand the current situation. This has been done through the recent studies and systems planning documents. From these a strategy for advancing can be developed.

It will be very important to develop long-range education plans for staff who will operate and maintain the organization's systems. Developing good staff is a long-term proposition. Staff should be selected based on their interest, previous education, and aptitude. Good candidates have some sort of natural science or geography background with an aptitude for spatial concepts—maps, orienteering, and so forth. At some point also consider tradeoffs between hiring contracting and training. Combinations to develop various strategies, such as contracting for complex and short-term tasks, contracting to fill while training occurring, or contracting to work with staff (on-the-job training)

Working in conjunction with the human resources department and participant staff, a specific education program can be developed as part of career planning, identifying specific education programs and training. Such career planning should be aligned with the requirements of applications.

Technology and Services Procurement

Implementation of the GIS will require new technology and specialized services including: hardware and software, networking and communications, management consulting, technical consulting, and perhaps various specialized data collection and conversion services.

The type and quantity of products and services required for GIS implementation will vary depending on many factors, including

- The business of the organization and the size or the organization
- The culture and philosophy of the organization with respect to contracting or undertaking work internal to the organization
- The scope of the project
- Data already in existence either within the organization or from another source (e.g., local, regional, or national government)
- The type and number of applications

Table 6.3 shows representative projects and proposes services that might be required for each type of project. Services and products that may be required, depending on circumstances, are shown with a hollow diamond. The products from these services will be required for implementation, but the organization may not contract for them because the products are already available, or because the work is performed by the organization. Those products and services that almost certainly will be required to be procured are identified with a solid diamond. Each product and service is briefly described in the paragraphs to follow. Although for convenience the products and services have been described separately and sorted into categories, it should be noted that frequently they are purchased in some combination, to create a package of products and services from a single vendor.

Consulting Services

This category refers to work performed by suppliers for clients, not including traditional geomatics[10] services, which are in the next category.

Management Consulting. Management and planning-related services include

- Identification of institutional arrangements, mission, strategies, and management plans—situational analysis
- User requirements analysis
- Feasibility studies
- Return on investment analysis
- Policy analysis
- Conceptual design and master planning
- Strategic planning
- Multistakeholder project design
- Change analysis and planning—business process redesign
- Staff analysis and recruitment
- Education and training seminars

Table 6.3   Products and Services Required for GIS Implementation

	Municipal Applications	Utility Applications	Natural Resources	Transportation	Commercial	Regional / National Gov't
Consulting Services:						
Management consulting	◇	◇	◇	◇	◇	◆
Project management	◇	◇	◇	◇	◇	◇
QA/QC		◆		◇		◇
Data conversion	◇	◆	◇	◇	◆	◆
Information tech consulting	◆	◆	◆	◆	◆	◆
Applications development	◇	◇	◆	◆	◆	◇
Geomatics Services:						
Cadastral mapping	◇	◇		◇		◇
Legal control & eng surveys	◇	◇	◇	◇		◇
Aerial photo	◇	◇	◇	◇		◇
Image analysis	◇		◇			◇
Photogrammetric mapping	◇	◇	◇	◇		◇
Project mapping	◇	◇	◇	◇	◇	◇
Hardware and Software:						
Hardware	◆	◆	◆	◆	◆	◆
CADD	◇	◆		◆		◆
GIS	◆	◆	◆	◆	◆	◆
AM/FM	◇	◆				
Image analysis			◇		◇	◆
GPS	◇	◆	◇	◇	◇	◇
SCADA	◇	◆		◇		
Docum't images	◆	◆	◇	◇	◇	◇
Multi-media	◇	◇	◇	◇	◇	◇
Raster scanning	◇	◇				◇

- Technology scoping
- Migration planning and other technology planning
- Technology requirements and procurement assistance
- Other related nontechnical consulting tasks

Project and Systems Management Services. These have the purposes of managing technical components of system design, construction, and/or operation, including

- System acquisition and installation
- Implementation of new applications
- Management of data conversion projects
- Selection and implementation of quality assurance procedures
- Management of technology migration from existing to new systems
- Facilities and operations management—outsourcing
- Related systems implementation and operation management tasks

Quality Assurance and Quality Control Services. These services include

- Data quality specifications
- Selection/design of data conversion procedures
- Design and management data loading/checking processes
- Determination of performance metrics
- System and application reviews (audits)
- Interpretation and application of total quality management standards (likely not for some time yet)
- Other related services (there may be considerable overlap between these and consulting services depending on how the project is structured)

Data Conversion Services. These can be provided at several levels and for many purposes. This category may have considerable overlap with traditional geomatics services or may have little relationship to them. Hence, data conversion has been singled out as a separate category under consulting services. The following are examples of data conversion services, but there are conceivably many others

- Conversion of manuscript maps to digital form—can vary from simple line following to produce digital map files or drawings by hand or using a scanner to enter coordinate geometry data to construction of complex spatial entities using any combination of the foregoing
- Conversion of engineering drawings, plans, as-builts
- Conversion of existing digital files from one format to another, including geocoding of tabular data sets, raster-to-vector conversions, feature extraction, and related formatting processes
- Key entry of attribute data
- Conversion of flat files to relational tables

- Conversion of documents, cards, or other manuscripts to document images
- Other combinations of manuscript and digital conversion techniques

Technical Consulting Services. These include

- Data analysis and modeling—entity relationships and data flows, enterprise models
- Database design—linkage of attribute data to spatial data, attribute database design, design of application specific and corporate databases
- Design of mapping and other technical specifications
- Determination of technology specifications—capacities, performance requirements, type and quantity of equipment
- Technology evaluation
- Network and other data communications design and implementation
- Technical training
- Related technical services

Applications Development Services. For the purposes of this discussion, these will include software development for specific purposes, assembling technology components into a comprehensive system, assembling data and technology to satisfy specific user needs, and decision support services wherein only the end product is provided. Thus this category could be further subdivided into several other categories depending on client needs and the business strategies of companies offering the services. The services included in this category are

- Design and coding of custom end-user applications for data retrieval, analysis, and decision support
- Assembly of technology components with or without data as determined by needs
- Assembly of data and technology—turnkey systems
- Decision support services wherein the end user receives only specifically tailored information products; may include presentation or viewing software as a means of delivering the results

Survey and Mapping Services

This category includes a broad range of services and information products associated with traditional surveying and mapping activities. There is some overlap between this and the preceding consulting services category, depending on the amount and level of management and technology services that are bundled with the geomatics services. Included in this category are

- *Cadastral mapping.* Capture and presentation of map elements that are created from legal descriptions of boundaries defining property ownership and other rights to land use—may include data conversion services or maps may be created as a result of new survey

- *Legal surveys.* Locating features and property boundaries applying measurements from a survey network (local or national geodetic network)—may include records research, monument recovery
- *Control surveys.* For mapping projects, construction, deformation monitoring, gas and oil wells and exploration activities—may include survey network analysis and/or design, vertical and horizontal control specifications, and monumentation (may involve the use of GPS technology)
- *Engineering surveys.* For construction projects, transportation, etc., involving many of the procedures listed previously
- *Aerial photo services.* Including planning and flying aerial photo surveys, compiling existing photos, and photo reproduction services
- *Image analysis.* Including acquiring, processing, and analyzing satellite and airborne imagery including Landsat, SPOT, and airborne multi-spectral data, radar data, weather satellite data, etc., and providing appropriately formatted data to the client
- *Photogrammetric mapping.* Using a stereoplotter or equivalent precision mapping technology, including the production of orthophotos
- *Project mapping.* Specific resource, land use, or other themes using any combination of mapping technology

GIS-Related Technology Products

These are technology products for the purposes of acquiring, storing, manipulating, and analyzing geographic and related data needed to support the identified GIS applications.

GIS Hardware. There is a huge variety of computer technology available for GIS applications. However, it can generally be categorized into the following:

- *Servers.* Data and computer processing servers in support of networks of user workstations and terminals
- *Workstations.* Consisting of personal computers (IBM compatible or Apple), and UNIX workstations (UNIX is the operating system most often used with these workstations—see discussion in Chapter 7).
- *Peripherals.* Digitizing tables, plotters, printers, film and image writers, and a growing list of others
- *Networking hardware.* To permit the linkage of various other devices

Computer-Assisted Design and Drafting (CADD). This is the appropriate technology for engineering design and digital mapping. Engineering design consists of CADD drawing construction using geometric operators. The result is a CADD drawing encapsulating the design parameters. The drawing (in physical and digital form) is then used for construction. This process is also sometimes referred to as CAD/CAM (computer-assisted design/computer-assisted mapping or manufacturing). A primary use of CADD technology in GIS projects is digital mapping, also called digital cartography (the making and study of maps), performed using a CADD system. The emphasis is on the graphic communication of spatial relationships and distributions. In current systems, it usually includes the analysis and manipulation of geographic data to enhance representation.

Geographic Information System. This is an information system designed to manipulate computerized maps and the geographically referenced information tied to those maps. GIS technology relies heavily on several other information system technologies including database management, computer graphics display, network communications, and distributed computing through high-speed networks. Contrasted with CADD, GIS combines graphic capabilities with strong nongraphic attribute linkages allowing complex query, map overlay, polygon processing, and spatial modeling operations.

AM/FM System. This is a collection of software, procedures, and policies to assist in the management of a geographically distributed facility. AM/FM may be considered an application of GIS technology, and most AM/FM technology includes features of GIS or is built upon a GIS product.

Image Analysis. This is information technology applied to the acquisition, manipulation, and analysis of remotely sensed digital data. Image analysis software has changed considerably in recent years and now often includes some GIS functions as well. Additionally, many GIS products now offer some capability to work with geographic images. The principal differences often reside with the internal data structure of the software.

Global Positioning System (GPS). A constellation of satellites originally developed by the U.S. Department of Defense as a navigational aid has become available for geodetic control surveying, location of features, and "on-site" digitizing. The satellites transmit signals that can be decoded by specially designed receivers to determine positions of varying accuracy (depending on the device and the existence of ground stations). These measurements can be made at a fraction of the cost of traditional first-order surveying techniques. Eventually the GPS network will consist of 21 active satellites and three spares, orbiting 10,900 nautical miles above Earth, that will allow 24-hour, all-weather operational capacity in both navigation and relative positioning. The U.S. Department of Defense is developing the GPS at a cost of more than $10 billion (United States Launches, 1992).

Supervisory Control and Data Acquisition (SCADA). This is technology based on remote instrumentation (telemechanics) technology for process control in office buildings, chemical plants, distribution networks such as gas distribution, electrical networks, and telecommunications networks. New applications of SCADA are developing rapidly, and SCADA technology is being merged with other systems to develop "real-time" GIS applications.

Document Imaging. This is information technology for conversion, storage, and retrieval of scanned documents. Documents are converted to high-resolution raster (grid cell) format that are stored as bitmaps that can be displayed on graphics devices. Documents appear on the screen as "photograph"-like images of the original manuscript. New document imaging technology includes features such as "zones," which allow parts of the document to be active and revisable, and "workflow," which facilitates documents to be routed through the organization in a predetermined manner.

Multimedia. This is a information technology application that facilitates merging static and active data types into a single database. Data types include: images, sound, animation, video, text, graphics and drawings, and with some limitations, real-time transmission of sound and images.

Raster Scanning. Raster scanning technology permits maps and drawings to be converted to digital format without the need for a human operator to identify each feature on the map and manually input the corresponding coordinates. However, raster scanners produce images of the map, not the vectors required for most mapping applications. Software is used to convert the image into vectors. There are many software products for converting from raster to vector, and this is by far the more complex part of the process. Raster-to-vector software is not fully capable of performing all the conversion tasks, and considerable operator intervention is still required.

Procurement Procedures and Policy

All organizations have procurement practices and policies. In large organizations, especially in government, these are highly formalized, may be governed by legislation, and may be managed by a separate procurement organization. Depending on the policies, practices, and culture associated with procurement, the existing system may be a source of problems for the GIS project. Some examples are:

1. The organization has adopted an information technology standard with which all technology purchase must comply, but GIS software with the required capabilities is not available for the organization's selected IT standard.
2. The organization has a policy of always selecting the bid with the lowest price.
3. Procurement policy only allows a tender or bid process excluding any

changes to specifications requested that might be proposed by the respondent.

4. The procurement process does not permit any negotiations with respondents.

5. The organization (likely government) has a "buy local" requirement, but there are not adequately experienced vendors available locally.

The preceding examples, while deliberately somewhat extreme, are representative of the types of issues that may be encountered when the organization first sets out to purchase GIS products and services. As more experience is gained, fewer problems will exist. In other organizations, procurement policy will not be an issue.

No matter what the situation is, two things are strongly advised: (1) that the GIS project team determine existing policies and practices and evaluate their impact (if any) on the procurement of products and services for the GIS; and (2) that the procurement administration staff (if there are any) be involved early in the process. It should be understood that existing policy was not developed to obstruct the GIS project. The existing policy was developed to support what at the time were valid requirements. Developing an understanding of why the policies were developed will help in the identification of an approach to satisfy both the policy and the needs of the GIS project.

Different procurement approaches best serve the needs for purchasing different services and products. The purchase of standard hardware items such as plotters and digitizing tables (commodities) may best be served through a tender process with rigid specifications and price as the major consideration; however, purchasing management consulting or database design services following a tender process may not produce the best results. In purchasing services that require these specialized skills, or creativity, or for which the precise end result of the service cannot be defined, a process that leaves more input to the supplier must be used. Such processes are usually referred to as *requests for proposal* (RFP). Many organizations that call their procurement documents and process and RFP are in fact still using a tender process (also known as *request for bids,* or RFB). In the RFB process, the specifications for deliverables are rigidly stated, and most of the decision is based on price. For an RFP process to work to best advantage, it should encourage creative solutions from the respondents, and the evaluation criteria should determine the best value, not the lowest price.

To undertake procurement in a manner that may fall outside normal practices or policies of the organization, it may be necessary to define and obtain *special dispensation* from some policies and practices. Such dispensation has been successfully received in some circumstances by identifying the GIS project as something special that would suffer as a result of enforcement of specific policies and practices. Dispensation does not alter existing organization policy, as it applies to other activities. It creates a special temporary administrative environment free from administrative barriers that might otherwise impede development. To be successful, the required dispensation must be well defined and justified and negotiated with appropriate authorities.

The Procurement Process

Depending on the policies and practices of the organization and on the issues discussed in the preceding section, the procurement process will vary somewhat by organization and the products and services being purchased. However, the process may generally use the steps to be outlined.[11]

Information Acquisition. Background information about available products and services is collected. This may be accomplished in several ways and in various combinations. A request for information (RFI) document may be prepared and widely circulated requesting prospective suppliers to send descriptions of their qualifications and products. Company representatives may travel to other organizations that are currently using GIS, or they may attend conferences. This process was described in Chapter 4 during the discussion of project startup. It may be repeated for the procurement process, or the information collected during project planning may suffice.

Functional Specifications. Functional specifications will have been developed during the needs analysis process. These will likely need to be confirmed and reformatted for use in the procurement process.

Request for Proposal. A document clearly presenting the needs of the organization, the requirements for responding (including due date list of clients, format of response, how it is to be delivered), the review process, the criteria used in evaluating proposals, and how respondents will be notified. If this document is truly an RFP and not a tender, the document should include a statement to the effect that "respondents are invited to present alternatives that do not precisely match the stated requirements." Such a statement creates the opportunity for suppliers to propose innovative alternatives that may not have occurred to the writers of the RFP. Some organizations may not entertain this option as it complicates the RFP process; however, there are many ways to satisfy requirements for both administration and innovation (see the Centra Gas example in the Appendix).

Proposal. The proposal document is prepared by prospective suppliers. The proposal evaluation process and selection criteria should be determined in advance of receiving the proposals and must not be changed after the proposals have been opened.

Proposal Evaluation. Antenucci (see note 11) proposes the following steps and actions for the evaluation process:

1. *Responsiveness review.* Does the document(s) comply generally with the requirements of the RFP, and in the format specified (if specified)?
2. *Ranking.* An initial ranking of documents that comply, sorting out the more promising responses from the ones that are not likely to be selected (e.g., the RFP requested hardware and software and the response only includes software).
3. *Rigorous review.* Applying the selection criteria, each RFP is evaluated for content, compliance with specifications, noting any special circumstances or innovative approaches to satisfying the organization's needs. Aspects of the process may include

   - Oral presentations by the respondents
   - Visits to the supplier's site (e.g., offices or technology development center)
   - Benchmarks—technical evaluations of performance and compliance with technical specifications by conducting rigorous performance tests on the software and hardware
   - Visits to user sites where the technology or services are currently in use
   - Reference checks—telephone interviews with others who have purchased the products or services to determine their satisfaction
   - Any combination of the above

4. *Cost review and reranking considering price.* Having determined the proposals that will satisfy the needs of the organization, re-evaluate the proposal for cost effectiveness (not lowest price). This is a complex issue because there can be hidden costs and risks, such as warranty disclaimers, support limitations, and so forth. Suppliers will often attempt to provide aggregate prices, allowing them to discount some parts of the overall system, to compensate for what they know to be higher prices for other parts. "Freebees" will often be thrown in to "sweeten the deal" but with limited support or guarantees. These may later be the source of significant expense. Protecting oneself from these issues is complex. If the system purchase is large and complex, involving a lawyer who specializes in information technology procurement contracts is a good idea.[12]
5. *Selection and recommendations to management.* Finally, having conducted thorough reviews, a recommendation to purchase will be made to someone, usually the management authority. Often the final approving authority has not participated in the review process and will receive only a recommendation. However, because they are the management authority, they will be accountable in the end for the decision. For these reasons (and others related to legal liability), it is important that the evaluation process be well documented and that the management authority is provided with a well-organized summary of the

evaluation process and results. Some procurements have become stalled as a result of management being uncertain about the recommended selection or the review process.

Benchmark Test. The benchmark[13] test was mentioned as an option in the evaluation process, but it warrants additional discussion. A benchmark test is a complex and expensive process, if performed correctly. It should only be used when cost and technical complexity warrant it. For a typical acquisition involving a dozen or more applications, a large database, and perhaps a hundred users, the cost to all parties of conducting the benchmark is conservatively $100,000. For the benchmark to be effective, the organization must create measurable specifications for each application and data management process to be tested. Benchmark procedures, a schedule, and an evaluation team must be assembled. Because the process is complex, many organizations will retain a consultant to assemble the benchmark test, further escalating costs. The vendor will spend days, perhaps weeks, preparing for the test and then spend additional thousands of dollars to bring staff and equipment to the benchmark site. The more money the supplier spends up front on winning the business, the less likely he will be to deal during negotiations.

Many GIS products are now in use in many sites. Good comparisons of performance requirements can be made by consulting those sites. The decision to undertake a benchmark test should only be made after careful consideration of the costs and alternatives. Requiring a benchmark when one is not warranted may exclude the best suppliers of the required products because they are unwilling to partake in what they consider to be an expensive, unnecessary process. It is also an expensive way for the organization to acquire its needed technology. A benchmark should not be considered as a substitute for a pilot project. There are more effective and less expensive learning environments if that is the objective.

Selection and Negotiation. Once the review is complete, the organization has accepted the proposal, and legal and administrative issues have been dealt with, it is time to notify the selected supplier(s). Again, depending on the policy and practices of the organization, final selection may involve a negotiation process. If not, then the supplier is notified and contract documents are prepared and signed.

If negotiation is to be part of the process, it should be identified in the RFP, indicating that a "short list" will be established for final negotiation. A negotiation process is more difficult to manage than a straightforward selection process, but it may yield better results depending on the purpose and objectives of the procurement. Playing one supplier off against another may result in a lower price or more for the money, but there are other considerations. The organization will have to work with the successful supplier for some time. Creating a poor relationship as a result of a one-sided negotiation process will likely result in a poor working relationship with the supplier. Negotiation should attempt to provide the best and most equitable deal for both parties, creating a partnership relationship with the supplier. Properly motivated, suppliers can contribute greatly to the success of the project.

Followup. Once the negotiations are complete, the unsuccessful respondents should be notified promptly. They should be provided with an opportunity for a debriefing, but decisions must be final. A proper, well-conducted debriefing will eliminate most attempts at overturning the decision. This stage in the process reinforces the value of a well-documented evaluation and selection process to deal with challenges to the decision.

Two case examples, Centra Gas, Winnipeg, Manitoba, and Manitoba Ministry of Forests, are given in the Appendix.

Implementing the GIS Applications

Planning activities have generated specifications and a general design. These must now be translated into design specifications that are adequately detailed to permit applications to be constructed. The activities of many staff, some external to the organization, must be scheduled and managed. Technical issues requiring attention are: preparation of detailed design specifications; data collection, compilation, and conversion; development of technology selection criteria; construction of the applications; and selection and installation of technology.

Previous sections in this chapter have presented the elements to be considered and managed in implementing GIS applications. In this section the parts will be brought together as a process, as diagrammed in Figure 6.4. The process in this figure assumes that several applications will be constructed incrementally. Toward the completion of each application, the implementation plan should be reviewed and updated, benefiting from what has been learned and scheduling resources and other participants for the next application.

The order in which tasks inside the shaded box are performed depends on the approach selected to conduct the work. For example, if a decision has been made to contract with outside companies for data conversion, then the work can begin after detailed design specifications have been prepared and the systems architecture (hardware and software) are known. If, on the other hand, data conversion is to be done internally, then the system must be installed and staff trained before data conversion can begin. Other tasks within the box are dependent on preceding tasks: testing and software customization.

The implementation tasks can be made more manageable with the aid of a methodology, as discussed in Chapter 5. It is important, however, to recognize that no two organizations, set of requirements, or management frameworks are exactly the same. The methodology selected is very dependent on results from the situational analysis. Having one and adapting it to the specific needs of the organization is more important than which is chosen. Beware of "cookbooks" or prescribed solutions to fit any situation: No two organizations are exactly the

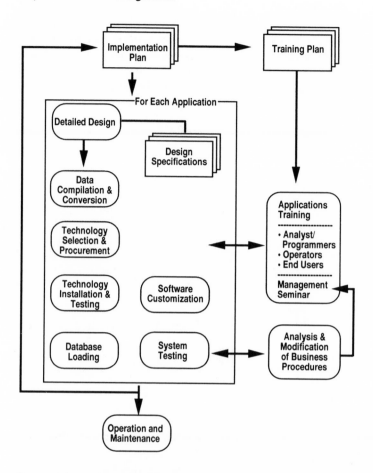

FIGURE 6.4 Detailed design and implementation

same. Failing to adapt a methodology to the needs of the organization may result in the "square peg in a round hole" syndrome, forced into a solution that does not recognize their unique needs or requirements.

The methodology should impose order on the development process, standardize activities, and enhance communication between members of the project team, including management and end users. An appropriate methodology should provide the means for practical partitioning of the complete system development process. The partitioning provides the basis for accurate estimation of costs and schedules, defining interim and final deliverables, and providing progress measures. The method must integrate the technical, managerial, and institutional issues in a framework that ensures high-quality deliverables. Results from the method should be readable and should provide management and auditors with confidence in what is being built and those building it.

The same is true for standards. Standard development should be restricted to those aspects where it is necessary. The developed standards should make the system more usable by end users, not inhibit or unnecessarily complicate their use of the system or data. New map symbology or other data standards should not be developed for the sake of the technology. Standards creation and coordination needs to involve all stakeholders in systems development and use. There are many areas in which standards may be required, more if the implementation involves participants from many agencies. A good review of standards issues is contained in Croswell (1990).

Detailed Design

The purpose of detailed design is to translate the generally stated requirements of the end users documented in the user (functional) specifications into a set system construction specifications. The approach selected to conduct the general design will now determine much of what is included in the detailed design. If the general design did not include the following, then the following must be performed now in conjunction with the detailed design:

- Comprehensive data and process models for each unit in the organization for which a GIS application will be developed
- Determining and structuring the contents of the geographic database to support the needs of all intended users, including data maintenance
- How and where data will be physically located and maintained
- The processes that will maintain the data
- Hardware performance requirements
- The type and location of equipment to perform required operations
- Communication requirements (networks)
- Software specifications for storage and processing of the data
- The user interface—the "look and feel" of the system, and how much can be performed by the user without the aid of a technician
- Documentation standards—what will be documented, by whom, and in what format
- Training needs
- Any special environment requirements, such as furnishing, lighting, climate control

As design proceeds, the process and data needs will define the overall system architecture and subsequently the technology selection criteria. If these were adequately determined during general design, then the technology procurement process can begin, dictated by the needs of each application. If general technology specifications were not developed during general design, then technology acquisition should not begin for any application until the overall system architecture can be defined to ensure that each application can be easily integrated to create a complete system.

Data Compilation and Conversion

The data required to support the applications must be determined, compiled and converted to a format usable by the GIS. The information requirements for each application will have been determined from analysis work performed during the needs analysis. However, the best sources or means of deriving the information may not have been determined during the needs analysis. Frequently there is more than one source of data from which to create the required information. Data may be available in an incompatible digital file, as hardcopy records, or from an external source. It is well worth the effort to undertake a small investigation of the costs and quality of the data from each source, prior to committing to conversion. For example, an incompatible digital file may exist in the organization, which at first look appears to be the best source. However, upon closer examination, it is determined that a specialist consultant at high cost would be required to convert the data. More important, it is determined that the file is 2 years old and has not been maintained. The external agency that collects and maintains the data can provide the data in a standard interchange format for a price. Although expensive, it turns out to be less expensive and more reliable than the alternative of converting the 2-year-old internal file.

In other instances, part of the required data may be available from one source and the balance from another. It may be that a 2-year-old digital file can be converted and updated with paper records at a lower cost than entirely converting the paper records. Or it may be that the digital records lack adequate reliability, that although they are an attractive source, upon closer examination they turn out to be useless.

There are two components to building the database required for GIS: initial input and maintenance. Initial input consists of acquiring by various means the data needed to support the applications. Maintenance consists of keeping those data current.

Information is central to all the organization's decision-making. Good data and data handling procedures are required to produce and deliver useful information for decision-making. It is broadly recognized that in GIS implementation, data-related activities are the main source of cost. It is also well accepted that significant savings can result from reducing the number of times a piece of information must be handled. Additionally, costly problems arise when different parts of the organization collect the same information inconsistently and then apply those different versions of the information to decision-making. Part of the focus for GIS is therefore to increase the efficiency of data handling and to make the right information more readily available for decision-making.

There are known approaches that, if adopted, can significantly reduce the cost, implementation time, and maintenance effort over the life of the system:

- Sharing data within the organization between departments and eliminating duplication in handing information—a common data model is key to facilitating such sharing

- Obtaining data from others—obtaining digital base maps, land records, and other basic data from government agencies and others
- Utilizing existing digital records
- Using alternative technology for data capture/storage—images, vector scanning, indexing (microfiche, paper files)
- Deriving data by using GIS processing—graphic overlay, relational matching, automated record matching, and topological overlay[14]
- Transaction processing—databases can be designed such that they are maintained by capturing records of the day-to-day activities of the company

Maintenance of data is a bigger issue than the collection of data because

- Data are of little value if they cannot be trusted to be current
- Construction of the database is a one-time effort, but maintenance is ongoing and therefore, over time, the cost of maintenance is higher and it is possible to obtain significant savings from gains in efficiency
- Changes in the database through maintenance procedures implies a need to maintain a history of past situations and the transactions that made the change necessary. This is technically complex and has not been adequately solved at this point, but it is likely to be costly even if performed efficiently
- There are many agencies involved in the maintenance of land records (for good reasons, as discussed later). However, this further complicates the situation and demands consistent means for dealing with data maintenance across all users
- Since there are many agencies involved in the process, there also exists a potential for significant redundancy that not only increases the cost but also leads to problems in maintaining consistency of the data across users

A useful rule of thumb at this stage in design and implementation is:

No data should be captured and converted without an associated plan for maintaining those data.

Technology Selection

An earlier section dealt with management and administrative issues related to procurement. From a design perspective, the key procurement issue is developing specifications and selection criteria that will ensure that the most appropriate technology is acquired. Design staff and users should participate in developing the specifications and selection criteria. Purchasing GIS technology is another one of the points in the process where it is important to gain commitment through participation. Besides, for most people buying GIS technology is the most exciting phase in a project—for many people it is akin to buying a new car.

It should not be necessary to collect new information at this point. All the necessary facts should be contained in various project documents (user specifications, general design documents). The emphasis will be on sorting out which are the most important features and ranking them accordingly. An example of selection criteria developed for the Alberta Planning Board project[15] is detailed in the Appendix.

Note that the criteria do not specifically identify the type or brand of technology, nor any other technical specifications. They are almost entirely related to use and to user measurable aspects. This approach places emphasis on the suppliers of the technology to assemble technology configurations that meet the defined needs. It also suggests that there are no existing technology standards to which one must adhere.

Another approach is to define specifically a *technology architecture*. An information technology architecture defines characteristics of the technology that will ensure that selected technology will satisfy the specified needs of user applications and ensure that the technology will work together as a complete system, without specifying brands of technology. Examples of characteristics that would define a technology architecture are

- Hardware performance specifications defining characteristics such as graphics display capability, processing capacity, storage media, memory capacity, adequate standard communication ports to support required peripherals (e.g., printers), and network communications capability. These are independent of a particular "brand" of technology. The main purpose of these standards is to ensure a consistent minimum level of performance for all users, that the hardware is capable of operating defined applications software, and that selected equipment can be interconnected creating a network of systems (as required).
- Operating system characteristics, such as the ability to multitask, support network communication, memory management, physical storage management, graphics display, and support required application software such as GIS.
- Communication protocol standards, which ensure that the selected technology will be capable of communication over a network.
- Application software specifications, which consist of minimum requirements for software that will be used to construct end-user applications such as GIS and database management. Examples of standards are: topology in a GIS and support of SQL (standard query language) in a database management software.
- Applications interface standard. This is the look and feel of the user interface for applications. A standard applications interface ensures that all applications developed for the GIS will appear the same to the user regardless of the software or hardware used to implement them. The

advantages of this relate to consistency of applications across departments and to subsequent training advantages.

The technology architecture should not state specific brands of technology, but should facilitate selection of the most appropriate compatible technology.

Technology Installation and Testing

Once again the scope and complexity of this series of tasks relate to the size and scope of the system. Testing a single workstation system is naturally not as complex as a network of workstations operating across a wide area network.

Technology testing should be planned in advance of receiving any equipment and software on site. Key aspects of the technology functions (software, hardware, and, if applicable, network components) should be identified from the procurement specifications. The procurement process should have resulted in selection of the best-suited technology. The purpose of this series of tasks is to ensure that what was specified has been delivered and that the functions operate as specified. This is different from the final system testing as will be described, which may include custom applications developed by the organization or third party.

The test plan should have been prepared as part of the implementation plan. The test plan should identify critical functions and data handling processes to be tested, equipment quantities and model numbers, who will perform the testing, who will "sign off," documentation requirements, schedules, facilities required and any other administrative details. The tests need not be performed in a single day, and it is not unusual to identify an evaluation period during which testing and signoff occurs. Final payment (usually in the order of 20 percent) is not made until testing is complete.

Database Loading

Completion of this series of tasks results in a working database containing all the data needed to support the GIS, or at least those applications identified for initial implementation. Database loading may be performed by the organization itself, or it may be contracted out.

If the work is to be performed by the organization, then installation of technology and staff training are precursors to a long series of data conversion tasks. If the data conversion and database loading have been performed by a third party contractor, or a systems integrator, then a test plan similar to that required for the technology will be required to test the database. Many organizations will contract for a third party consultant to provide data testing and quality assurance, bringing to the process extensive experience and often custom quality assurance software that can find problems.

Software Customization

Most GIS software provides a set of tools and a development environment for building user-specific applications. Therefore, most organizations should plan on conducting additional software customization. If significant amounts of customization are required, then one of the following options is suggested:

1. Purchase customization as part of the GIS acquisition, paying the vendor to develop the required applications. If there is a larger market for the application, it may be possible to negotiate a price reduction in exchange for the rights to market the application to other organizations.
2. Contract a reputable third party developer, such as a systems integrator with experience with the selected GIS product.
3. Plan internal customization well in advance of the requirements. Plan for extensive staff training or recruitment of qualified personnel. Ensure that the amount of customization has been estimated in advance, that it has been budgeted for, and that project sponsors and end users are fully aware of the amount of work required.

All three of these alternatives are known to have been successful in different organizations. Which is selected depends on the characteristics of the organization as discussed in Chapter 3.

System Testing

When an application is completed there is a need to sign off on the development. This should be marked by a system test similar to that conducted when the technology was received.

A test plan should be developed as part of the functional specifications, indicating performance measures for the application (how will it be measured whether the application works or not). The test plan and performance measures should not be left until completion of the application because

- In the absence of clear performance measures, development staff will not have clear guidelines to ensure accurate development
- User expectations will change over time

The latter point is difficult and explains the insistence on end-user participation in all aspects of GIS planning and design. It is normal and mainly positive that users continue to learn. However, ever-changing specifications equate to never-finished systems. A balance between these two perspectives needs to be struck. The purpose of the performance measures is therefore to serve as a record of the expectations at the time of the design as a baseline for measuring progress and not as a set of regulations to be adhered to at any cost. Each change to the functions of an application should be reviewed in the same context as the original specifications, and any modifications to the application scheduled in relation to other priorities. Such an approach creates an open, innovative, but managed development environment. When a great deal of uncertainty exists, the best way to proceed may be through a pilot project or prototype.

Pilot Projects and Prototypes

As discussed in Chapter 4, pilot projects can serve a variety of purposes, including building understanding (training), reducing uncertainty and risk, and determining impact on operations and procedures. *Pilot projects* are raised again in this discussion because they can form an integral part of the implementation process, even if they were not explicitly planned for at the outset. They can take many forms: We are using the term in a broad sense to include tests lasting only a few weeks to significant projects lasting many months. Pilot projects are frequently defined as temporary working environments in which costs, technology, staff, and organizational impact can all be controlled. They create an environment where failure (i.e., something not working as predicted) can be tolerated.

If pilot projects are added as an afterthought, it should be kept in mind that they require staff time for planning, data collection, performing the pilot, and evaluation. This will impact existing production schedules and should be approved by the management authority. The scope, number, and formality of pilot projects can vary widely. Pilots should always have a specific stated purpose, start, and end. They cannot be substitutes for careful design and planning. However, they can be good supplements to training and can provide project staff with an "off-line" learning environment. Evaluation of results should be thorough and involve participants from all affected parts of the organization.

Another widely accepted technique used in implementing applications is *prototyping*. A prototype can be defined as a working model of the operational system. The advantage is that the rigor required to create a reliable operational application is not built in. Often a computerized prototyping tool or similar technique will be used to build the prototype, emphasizing the user interface, to mimic the look and feel of the application. The underlying functions and data structures will not have been built. This can be very deceiving for an end user and care must be taken to make the limitations of these techniques clear if either technique is used. Neither technique will have incorporated the thoroughness required for an operational system. If it had, it would not be a pilot or a prototype.

Demonstrations and Presentations

It is very easy to get wrapped up in work and to become familiar with technology. However, others in the organization will not, and they will want to see what the GIS can provide. Demonstrations and presentations perform an important role. They will be more successful if they are planned for and have had adequate time and resources allocated.

Project Reporting

Good communication is closely related to success. A successful implementation requires keeping all participants, and especially project sponsors, informed of progress. In project management "no news is not good news." Good reporting should begin by considering several factors: Who needs to be communicated with, how often, and for what purpose, or, more specifically:

1. Who is the report for and what do they need to know?
2. How can this information best be conveyed?
3. How can the burden on technical staff be minimized?

Project sponsors and senior management generally want general information on project progress at prescribed intervals. The reports should be brief and to the point. In many organizations, management reporting deals with exception reporting, that is, those aspects of the project that are not proceeding as planned, what is being done about it, and when the problem will be corrected. The exact content and frequency of management reporting depends on the culture and practices of the specific organization.

There are many others who may need to be communicated with:

- End users and other participants in the project
- Nonparticipating employees of the organization
- Others, outside the organization, such as data providers or users, or others in a similar industry who may have professional interest

The amount of effort expended on other forms of reporting depends on need and available resources. Many GIS projects in recent years have produced multipurpose newsletters to report on progress and new developments. For internal staff communication, monthly lunchtime talks or afternoon seminars, perhaps including a demonstration or prototype, may be the most successful communication tool.

Closing Remarks on Project Management

Most project-management-related issues have been discussed in previous sections. However, there are few topics that have not been covered in preceding sections. These have been presented here as tips and anecdotes.

Much of project management relates to people and getting the most from them. Here are several tips on how to get the most from staff:

1. The demands of the task must relate to the skill of the person assigned to the task.
2. Assign individual related tasks.
3. Assign critical tasks to the most reliable people.
4. Assign complex tasks to the most reliable people.
5. Assign tasks that interact with the same person (or with two people who can communicate).
6. A dedicated, less qualified individual is preferred to a person who is overworked or who has multiple priorities.
7. Offer the opportunity for growth.

There are five steps in a project, and they can be represented as two loops.

Planning Tasks (Triggers). (Discussed in Chapter 4.) These are to

- Establish the objective(s)
- Plan the project and its component parts

The project is generally approved after these two steps. There will often be a hiatus at this point while the necessary infrastructure or changes are lined up.

Implementation Tasks (Bullets). These are to

- Implement the plan
- Monitor feedback
- Adjust the plan implementation as necessary

Too often the North American/European cultures have very fast triggers and very slow bullets. This is in great contrast to Japanese. As a general rule, slowing the trigger greatly improves the chance of success for the project.

Project management entails planning what is to happen and providing controls to see that it does. Projects usually wander off track gradually, like slow erosion. Beware of "creeping incrementalism," wherein the project scope grows by small increments until it can no longer be completed with the time and resources available. This represents an erosion of resources—time, funds, staff—where a little at a time are siphoned off for non-mission-critical activities. Most missed targets do not happen all at once: They happen a little at a time.

Practice saying no. Stick to plans and schedules. Provide mechanisms for dealing with the addons without compromising the big result. If any milestone is missed by a significant amount, the project will not be completed on schedule.

You can "sucker" a project manager into accepting an impossible deadline, but you cannot make him or her achieve it. Countless disasters are caused because people allow themselves to be pressured into committing to something they felt was impossible.

When problems occur and deadlines are missed, explain what happened, but do not make excuses. Explain what is being done to correct the problem, state when the target will now be achieved, and do it. Do not return several times asking for more time and resources to solve the same problem. Therefore, do not underestimate a second time and have to come back for more time or funds. Do it once; do it right. Every project has problems, and even good managers miss targets once in a while; the difference lies in how they handle them.

More serious problems may arise from shortcomings with the plans, process, or specifications. For various reasons planning may not have been as thorough as it should have been, and many details may not have been fully determined. The design specifications and the cost and time estimates based on them may be found lacking. Sociological and institutional problems during design may have created animosity with some users or their management. These may result from communication problems during system design, resistance to technological change from perceived personal threats to job or power, lack of executive commitment, underestimates of required resources, overestimates of returns, inadequate integration of the system with business functions, and a lack of will or ability to make the institutional changes required to derive cost effectiveness from the system.

It is essential that a management framework that can deal with these and other issues be in place. The project and business managers must establish and maintain credibility with the management authority (steering committee, executive committee, or equivalent) and that issues can be raised and dealt with collectively and openly. There will be problems, but successfully resolving problems is essential for successful implementation.

NOTES

1. As a guideline, Dickinson (1989) suggests the following percentages of effort for each major development activity (excluding system installation, which can vary greatly depending on the individual environment):

Initiate project	up to 2 percent
Preliminary analysis and design	about 8 percent
Detailed analysis and design	about 65 percent
Build system	up to 25 percent

2. Figure 6.1 is a modified version of the process presented in Figure 4.1. The model now includes the added planning and management activities.

3. This issue was discussed in a variety of contexts in Chapter 2, including GIS principle 8: A GIS causes changes in procedures, operations, and institutional arrangements among all users.

4. Chapter 1 discusses in detail the paradigm shift necessitated by the implementation of GIS technology: the demand for changes to the present business processes and the resulting new way of looking at the functions of the organization. Because all aspects of the organization are eventually impacted, successful implementation must deal with all aspects.

5. The section entitled "Building Project Teams" in Chapter 4 describes the significant differences between commitment and compliance and the resulting impact on an implementation project.

6. Motivating staff and creating environments conducive to innovation and active participation is a complex subject. In addition to the issues raised in the text, different people behave differently. An excellent discussion of this topic can be found in Lane and DisTefano (1988).

7. According to Burg (1992), "Although only two per cent of North American companies outsourced their information systems (IS) facilities and staffs to third-party vendors in 1989, we predict that 20 per cent will follow suit in 1992. There are many reasons why the benefits of outsourcing cannot be ignored.

Consider the following:

- Outsourcing typically shifts many IS expenses from the capital budget to the operating budget. This can have a significant potential benefit in cash flows since capital assets become the responsibility of the outsourcing vendor.
- While outsourcing can save IS department budgets 10 or 20 per cent or even more in the first year, the power of compound interest will, as in your savings account, make even more modest savings monumental.
- For an organization contemplating the building of an information center from scratch, outsourcing will save millions of dollars immediately just in hardware and hiring costs.
- With economic growth at a virtual standstill, organizations are reluctant to make the capital investments necessary to update IS resources. And yet, they still want the strategic advantages of upgraded IS. Outsourcing is an alternative solution that can meet fiscal and strategic considerations" (p. 24).

See also the Grand Rapids, Michigan, example in Kraemer et al. (1989, pp. 76–77).

8. In February 1991, the Ontario government announced the creation of Teranet Land Information Services, Inc., a new company jointly owned by the Government of Ontario and Real Data Ontario, Inc. (a consortium of Ontario geomatics and information technology firms). The Province of Ontario Land Registration Information System (POLARIS) forms the basis for the Teranet Agreement. POLARIS consists of two digital data sets, the title index and the property mapping. Together these two data sets form the basis for the property cadastre for the province. This information is essential for the subsequent development of all municipal land information systems in the Province. Teranet is being run as a for-profit business, as reflected in the fees for the data. Consequently, the project is very controversial within the Ontario geomatics community, and some municipalities are undertaking their own property mapping in protest of the high fees.

9. At least at the time of writing—in the future the required skills may include object-oriented database design and development.

10. The Geomatics Industry Association of Canada and the Canadian Institute of Geomatics (formerly the Canadian Institute of Surveys and Mapping) define *geomatics* as "those disciplines involved in the application of sensing, mensuration, computer and communication technologies to acquire, analyze and manage spatially referenced information." For practical purposes, the field of geomatics is often delimited to include: surveying, mapping, remote sensing, activities related to developing and using geographic information systems, consulting, and associated education/training activities. The section

"Land-Related Sciences" in Chapter 2 includes a more detailed discussion of what the term *geomatics* includes.

11. Much of the following discussion of the procurement process has been derived from notes of a lecture on procurement by John Antenucci, PlanGraphics, in the course "GIS Project Design and Management" at the Banff Center for Management, October 1992. While the ideas are Antenucci's, their interpretation is ours, and the authors are responsible for any misquotes or errors.

12. For a good overview of the issues, see Takach (1989).

13. Aronoff (1989) describes benchmarking as the administration of a standardized test of a procedure to provide a systematic means of comparing the performance level of competing systems. In this way, the purchaser can better assess how a system will perform under the expected operating conditions. The evaluation should start with data that the vendor have not seen before. These data should be samples of real data sets that will actually be used in the GIS. The evaluation should include every major processing function and output product that will be needed and should be performed without stopping to modify the software. The system should perform without requiring software specialists to coax it along. This type of evaluation, performed under observation, will highlight unexpected difficulties and demonstrate the flexibility and robustness of the system.

14. These topics are not covered in this book and have been included as examples only. If the reader is not familiar with these terms, then a GIS professional or technical text should be consulted.

15. The Alberta Planning Board (APB) conducted a 24-month pilot project to investigate the use of land information and geographic information systems (LIS/GIS) for municipal and regional planning needs. The objectives of the project were to bring in affordable technology and resolve the technical problems and institutional issues that inhibit the use of GIS in small municipalities. The focus was on technology *transfer* and *institutional issues*. A distinctive aspect of this project was multiparticipant coordination, which reduces costs, redundancies, and obstacles. Technical requirements that must be satisfied for successful implementation were found to include

- The definition and creation of standard spatial entities such as properties, road segments, and utility infrastructure
- A communications network encompassing procedures and standards to permit electronic document interchange
- The development and adoption of data and related standards to permit implementation of a common data model throughout the province
- Mechanisms for successfully resolving stands issues, to be developed in conjunction with the Provincial LRIS Project. (Alberta Planning Board LRIS Pilot Project Management Committee, 1991)

REFERENCES

Alberta Planning Board LRIS Pilot Project Management Committee (1991). Final report.

Aronoff, S. (1989). *Geographic Information Systems: A Management Perspective*. WDL Publications, Ottawa.

Berg, Monica (1992). Outsourcing—an alternative solution. *Canadian Information Processing*, March–April.

Croswell, P., and A. Ahner (1990). Computing standards and GIS: A tutorial. In *Papers from the 1990 Annual Conference of the Urban and Regional Information Systems Association,* vol. 2, pp. 88–105.

Dickinson, Brian (1989). *Developing Quality Systems: A Methodology Using Structure Techniques,* 2nd ed. McGraw-Hill Software Engineering Series. McGraw-Hill, New York.

Dickson, P. R., and J. K. Simmons (1970). Behavioral side of MIS: Some aspects of people problems. *Business Horizon* 13, no. 4:59–71.

Hersey, Paul, and Ken Blanchard (1982). *Management of Organizational Behavior: Utilizing Human Resources,* 4th ed. Prentice-Hall, Englewood Cliffs, N.J.

Kraemer, K., J. King, D. Dunkel, and J. Lane (1989). *Managing Information Systems: Change and Control in Organizational Computing.* Jossey-Bass, San Francisco.

Lane, Henry W., and Joseph J. DisTefano (1988). *International Management Behavior: From Policy to Practice.* Nelson Canada, Scarborough, Ont.

Metzger, Philip W. (1981). *Managing a Programming Project,* 2nd ed. Prentice-Hall, Englewood Cliffs, N.J.

Takach, George S. (1989). *Contracting for Computers: A Practical Guide to Negotiating Effective Contracts for the Acquisition of Computer Systems and Related Services.* McGraw-Hill Ryerson, Scarborough, Ont.

Townsend, R. (1970). *Up the Organization.* Knopf, New York.

United States launches 17th GPS satellite (1992). *GIS World* 5, no. 4:10.

7

Managing the System

There will be a time when the implementation process ends and the system becomes operational. Exactly when this occurs may not be same for all projects: It may happen when the base map is finally completed, either through data conversion or by transfer from another agency; it may be when the first major application is finally operational and under control by the user; or it may be when an audit is done to determine where all that money went. Whatever the event that changes the project's status from a new idea to an operating system, managing the work will change from activities related to planning, educating, and designing to the more repetitive activities related to budgeting, managing projects, managing people, supporting the technology, and delivering products to the users. This is the time when the "GIS champion" becomes the "GIS manager"—a person who can manage an office of technical professionals and information systems technology to deliver consistent, high-quality geographic information systems support to the users. This is the time when promises must become reality—when attention turns to the operation and maintenance of the system.

Operating and maintaining a GIS is the same as managing a traditional information system in an organization. It requires an organizational placement, qualified and productive personnel, a means to manage the work, and an annual budget that can support the activities of the unit on an ongoing basis. This chapter investigates these important topics, for when the GIS project becomes an operational system, its continued success depends upon how the resources are managed on an ongoing basis to support the needs of the users.

Organizational Placement

One of the earliest and most perplexing issues to be addressed in managing a GIS project is where to place the responsibility for the GIS unit within the organization. While the idea, and quite possibly the GIS champion, may have come from one of the existing functional units of the organization, the enterprise-wide nature of the system implies that the responsibility for the system should rest on an organizational unit that has an enterprise-wide focus: the office of the top management official, the central administrative office such as a department of administration, or the central data processing unit of the organization. Some successful GIS implementations, however, have placed the responsibility for the system within an operating unit of the organization (the planning department, the public works department, the tax assessor). It is not clear why this is the case, but one explanation that has been offered is that GIS technology involves more than the traditional database and application development activities inherent to traditional systems that do not involve the complex cartographic and topographic concepts that are found in GIS technology. Professional data processing experts are not trained in these complex geographic skills, and therefore additional skills must be included in the project team (Ventura et al., 1992). Another explanation may be that GIS implementation is following the same path that the broader IS field took several decades ago when nearly all IS organizations were located in the finance department. Over the years, IS began to move out of finance and into other functional departments, general administrative units, or separate independent units. By 1985, however, only about half of the IS organizations in U.S. local governments had moved out of the finance department.

Experience tells us that there is no one ideal organizational placement for the GIS responsibility since each organization has different personalities and different organizational cultures. What is clear, however, is that successful GIS implementations are usually placed in an organization in a manner that reflects the structure of the larger organization as a whole. That is, if the top elected official is one who is a "hands-on" manager with interests in the day-to-day operations of the many facets of the organization, then the responsibility for the system is likely to be directly under that person's control in a centralized fashion. If, however, the top elected official is one who prefers to rely on departmental managers to deliver daily services and not to get involved with short-term management issues, then the GIS function is likely to be decentralized and under the control of some other department or body of departments.

The three most likely alternatives for placing the GIS responsibility in an organization are: within an existing operational department (possibly where the GIS champion is located); as a separate function within an enterprise-wide administrative unit; or as an independent unit directly under the control of the top elected official. Each alternative can work successfully, depending upon the personalities of those involved and depending upon the organizational structure and management culture of the organization as a whole.

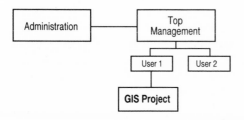

FIGURE 7.1 Operational organizational placement.

Operational Organizational Placement

A GIS under the management and control of an operating unit of the organization (Figure 7.1) can be successful if the manager of the operational unit has an expanded view of the GIS in the organization. This is a tenuous arrangement, however, because it depends entirely upon the personalities of the managers of the functional unit, and it requires a stable environment without emergencies or unforeseen issues. This is because when the responsibility for the GIS resides within an existing functional unit of the organization, it has primary responsibility to that organizational unit. Even if the system has been designed and planned for an enterprise-wide audience of users, the management of the day-to-day activities of the unit are still under control of the managers of that unit and, therefore, becomes directed by the mission of the unit.

Take, for example, an extreme case where the GIS responsibility resides within the police department of a local government. If a current violent crime wave were to occur and be widely publicized in the media, you can be sure that *all* resources of the department would be made available to address the problem, including the GIS. If, at the same time, a new solid waste recycling program were to be initiated by the public works department that could also use GIS capabilities for planning and management, which application would receive the most attention by the GIS team? Of course, a serious violent crime wave merits all the resources it can use, but the Police Chief has no responsibility for the recycling program, so where should he or she place his or her priorities? Clearly, the mission of the department responsible for the GIS will drive the priorities for its use.

Administrative Organizational Placement

A GIS under the management and control of an administrative function of the organization (such as a department of administration, a data processing department, or other enterprise-wide organizational unit), as depicted in Figure 7.2, can assure a balanced and equitable use of GIS resources across the organization. Such an administrative placement is usually under the management and control of a committee of user managers: a steering committee. Whether the steering committee sets policy and approves major projects proposed by the GIS

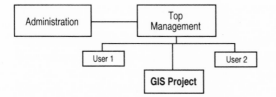

FIGURE 7.2 Skill/service organizational placement.

project team, or whether a strong manager such as the data processing manager sets policy and influences major projects, the results of this organizational placement of the system are most likely to be expensive and involve complicated technology. As described in Chapter 3, the Kraemer et al. research (Kraemer et al., 1989) has described this organizational placement as either the skill state or the service state of computing management. The common features of these states are: computing staff and budget are high, user charges are subsidized by the general fund, applications are oriented to support the functions of influential departments, and the chosen technology is either state of the art or consists of a set of diverse vendor products. The emphasis of applications for GIS utilization under this organizational placement will be either the influential members of the steering committee or the interests of the manager of the system. This structure reflects a common top management view that computing is largely a service function and therefore should be managed in a similar fashion as the personnel function, the purchasing function, or accounting.

Strategic Organizational Placement

A GIS in a strategic placement in the organization is under the direct control of the top elected official (Figure 7.3). This is the most political environment for the GIS project because policies for GIS are set at the highest level of the organization. The GIS manager reports directly (or indirectly) to the top elected official of the organization, placing him or her in the position of responding to the strategic needs of the entire organization.

Again from Chapter 3, Kraemer et al. (1989) indicate that this "strategic state" of organizational placement ensures that GIS applications are focused on public services and areas in favor of top management. It precludes the horizontal diffusion of computing capabilities to the operating departments, which could

FIGURE 7.3 Strategic organizational placement.

weaken top management control. Also, by regulating costs to user departments, this organizational placement is the most effective at keeping GIS staff and budgets low (after all, it is the top elected official who is responsible for the budget of the organization). However, the agendas, ambitions, and interests of top managers will influence decisions regarding the use of the GIS.

And so it is with all possible organizational placements of the GIS responsibility: Personalities make the difference. Whether it is the functional manager, the IS manager, the GIS project director, a body of influential members of the GIS steering committee, or the top elected official who is directing the project, the ambitions, responsibilities, and personalities of those in control set the pace for GIS services in the organization.

Kraemer et al. (1989) found that the responsibility for the IS organization in the local governments they studied reflected the role and status of IS as prescribed by top management. They found three aspects of organizational structure that influenced the IS role and status:

1. The fit between the organizational structure of the IS organization and the parent organization was generally tight. Decentralized local governments usually have decentralized computing and vice versa.
2. The departmental location of the IS organization similarly reflects management decision and action. While the MIS literature prescribes locating the IS function in an independent department on the grounds that such structure gives IS equal footing with other functions, top managers in most local governments have chosen not to follow this advice because of their own beliefs on how the function should be managed and controlled in their own administration.
3. The IS position in the organizational hierarchy is a significant indicator of top management's view of the IS function and of its commitment to computerization. Over the years, as IS grew in importance to the organization, it gradually moved up in the organization hierarchy in order to accommodate its increasingly broad organizational role.

Managing GIS Personnel

As Joseph Izzo (1987) writes,

> There is no such thing as a typical "computer person" and most seem to pride themselves on that fact. More often than not they think of themselves as creative types, as seen in their dress, diet, irregular hours, and loner attitude. As a result, they often regard themselves as different from other employees, somehow outside the normal rules and conventions that govern the business. This, needless to say, has created resentment and frustration throughout the firm. (pp. 30–31)

He goes on to explain, however, that this "high priesthood" of the information systems professional has been exacerbated by the lack of structure, standards, and procedures in the IS work environment. Without consistent practices and procedures, they begin to think that they do not have to follow any rules and can do their own thing. That, Izzo concludes, does not contribute to a spirit of professionalism, leading, in fact, to a department full of "nonresponsive prima donnas." He calls for the establishment of a more professional attitude on the part of information systems management to eliminate conditions that allow these employees to think that they are a special breed.

The professional attitude that IS (and thus, GIS) managers must establish requires a structure much the same as some of the more mature professions such as engineering and accounting. It involves clearly defined job descriptions, performance evaluation criteria, and associated career paths. It requires a consistent methodology for assigning and managing work. And, because of the complexity and technically demanding nature of the work, it must provide a working environment that allows a certain amount of autonomy over their work, allowing some degree of experimentation and flexibility in failing when attempting to try something new. These issues will be addressed in more detail after first exploring what the emerging "GIS profession" is and what the staffing needs are in implementing and operating a GIS operation.

Defining the GIS Profession

The expansion of the GIS industry and its projected growth rate estimates of 20, 30, and 40 percent for the 1990s and beyond have prompted a number of industry surveys seeking information on what the needs for skilled and experienced human resources will be to support these systems and satisfy the information hunger of their users. The results of these surveys have begun to develop a sense among the professionals and their employers about what this emerging GIS profession is.

The ESRI Study. A 1989 survey by the Environmental Systems Research Institute (ESRI 1989) collected information about staff sizes and job descriptions from their customers who attended its annual User Conference in May 1989. The results, published in the Summer 1989 edition of Arc News, showed that state and province agencies appeared to have more staff devoted to GIS and also had the highest percentage of job descriptions specifically related to GIS technology than either local or federal agencies (Tables 7.1 and 7.2).[1]

The Milwaukee and South Carolina Surveys. In Chapter 4, nine diverse roles of GIS professionals were described for staffing the GIS project team: manager, database administrator, analyst, cartographer, system administrator, drafter, programmer, digitizer, and processor. For the most part, early GIS professionals with these roles came from backgrounds in diverse disciplines such as geography, anthropology, engineering, surveying, planning, computer science, and related fields. Early job announcements also reflected a diverse background

Table 7.1   Staff and Job Descriptions by Level of Goverment

Organization Type	Average Number of Staff	Median Number of Staff	Percent with GIS-related Descriptions
Local	6.2	5.0	19
State/province	9.0	7.5	69
Federal	5.3	5.0	44

Source: Survey of ARC/INFO users from ARC News. Summer, 1989.

in their requirements for vacant positions, citing cartography, computer science, engineering, geography, information management, and urban planning. Often, the phrase "or related field" was added to ensure that enough applicants might apply for these vacant positions.

Two 1990 surveys, one by the City of Milwaukee and the other by the State of South Carolina Water Resource Commission, sought information about existing job titles, education and experience background, and salary ranges from agencies (including, in the South Carolina study, private companies) with existing geographic information systems. Both surveys found that not all agencies had the same job descriptions, or even the same responsibilities. The educational requirements for the positions ran the gamut of related disciplines identified in the previous paragraph. Of the 26 agencies that responded, 5 of the GIS roles listed were identified :

- *Manager.* Eighty-eight percent of the agencies reported positions with responsibilities similar to the GIS manager. Job titles included GIS

Table 7.2   Staff and Job Descriptions by Type of Position

Role	Sites Having Staff to Perform This Role (%)	Filled Positions Having GIS-related Job Descriptions (%)
Manager	90	33
Senior processor	90	66
Database administrator	76	54
GIS analyst	74	56
Digitizer	63	53
Programmer	53	36
System administrator	51	42
Junior processor	36	46
Cartographer	26	36

Source: Survey of ARC/INFO users from ARC News, Summer,1989.

coordinator, data processing manager, GIS project manager, principal programmer analyst, computer mapping manager, manager of GIS, computer graphics manager, and GIS department manager. Educational requirements varied from no degree specified to a B.S. degree in computer science, cartography, or engineering to an M.S. degree in geography or closely related field. Minimum experience required ranged from 2 to 5 years, most also requiring supervisory experience.

- *Analyst.* Eighty-one percent of the agencies reported positions with responsibilities similar to the GIS analyst. Job titles included senior systems engineer, systems analyst–senior, MIS supervisor, FM applications analyst, applications systems supervisor, senior systems analyst, and GIS systems analyst. Educational requirements varied from a B.S. degree in information management, geography, urban planning, computer science engineering, or cartography to an M.S. degree in geography, computer science, or closely related field. Minimum experience required ranged from 3 to 5 years.
- *Processor.* Sixty-nine percent of the agencies reported positions similar to that of the GIS processor. Job titles included database analyst, programmer analyst (GIS), lead programmer analyst, and GIS analyst. Educational requirements varied from a B.S. degree in geography or computer science to a B.S. degree in information management, geography, urban planning, computer science, engineering, or cartography to an M.S. degree in geography. Minimum experience required ranged from 1 to 2 years.
- *Programmer.* Only 19 percent of the agencies reported positions similar to that of the GIS programmer. Job titles included applications designer, intermediate data processing programmer analyst, applications programmer, and computer programmer II. Educational requirements generally included any B.S. degree. Minimum experience required ranged from 1 to 2 years.
- *Digitizer.* Forty-six percent of the agencies reported positions similar to that of the digitizer. Job titles included automated mapping operator and GIS technician II, and no college degree was required. This is the only position that reported no experience was required.

The results of these surveys indicate that there are no standard job titles for GIS professionals and that only about half of the people currently performing GIS-related functions in government were credited with having unique jobs (Huxhold, 1991).

The State of South Carolina survey was undertaken to propose a plan for standard job classifications and pay ranges across the 11 state agencies with active GIS programs. This, it was felt, would enhance recruitment efforts, provide career paths and salary advancement opportunities, maintain equity among all GIS professionals in the agencies, and accurately reflect the work of these professionals. The effort has led Byrd and Hale (1991) to develop the

**Table 7.3   Recommendations for Getting GIS
                    Positions in Your Organization**

- Work with your personnel/human resources department
- Aggregate the number of positions (classifications) currently used for GIS personnel
- Conduct a salary survey (anonymously within your organization and among similar organizations)
- Document any emplyment trends in your organization's ability to recruit or retain employees due to salary/compensation issues
- Document the unique employment characteristics of the GIS professional (multi-disiciplinary) nature of expertise
- Document the growth in the industry in addition to your organization's investment in the technology
- Consider designing career ladders if traditional means of salary advancement are not available
- Recognize the differences between the concepts of job classification and compensation
- Make realistic recommendations

Source: State of South Carolina (Byrd and Hale, 1991).

recommendations given in Table 7.3 for other agencies that are about to create GIS positions in their organization.

The Geomatics Industry Association of Canada Survey. In addition to nonstandard job descriptions, the Milwaukee and South Carolina surveys indicate that GIS experience and education may be more important for landing a job than is the type of degree possessed by the candidate (Huxhold, 1991). This situation reflects the relative immaturity of the GIS profession, especially in terms of formal educational opportunities in North American academic institutions. While over 400 universities currently offer at least one course in GIS (Morgan and Fleury, 1992), to date none offers a formal degree program for the profession.[2] This lack of a standard and formal educational program for GIS professionals prompted the Geomatics Industry Association of Canada in 1990 to join forces with the Canadian Institute of Surveying and Mapping to conduct a Human Resources Planning Study, which surveyed private industry, government agencies, and public companies to summarize the industry's concerns about human resources for the GIS (geomatics, as its known in Canada) industry. While the survey concluded that the number of GIS-related jobs in Canada increased by about 33 percent during the period from 1985 to 1990 and projected a 25 percent growth rate for "the next decade," 81 percent of the firms surveyed said they had some difficulty in hiring personnel. Almost half expressed concern that the educational system will not be able to produce the graduates they will need: "It documented industry concerns that measurement and data collection were being given greater weight in educational curricula than spatial analysis, information management, and GIS applications" (Forrest, 1993). "Measurement and data collection," as it relates to GIS, is generally taught in surveying, remote sensing, and civil engineering curricula; spatial analysis in geography; information management in MIS programs of business or computer science programs; and GIS applications are taught in almost any discipline that studies physical or socioeconomic phenomena on the earth (or on any globe, for that matter). What this survey tells us is that skills from many different disciplines are needed to develop effective and marketable GIS professionals—the highly technical engineering-related degrees are not the answer. What courses, then, from what disciplines are needed to produce the GIS professional?

The Oklahoma State University Survey. The answer to the previous question may have some basis in the results of a survey of GIS industry agencies that was conducted by Oklahoma State University in 1990 in an effort to design a curriculum that can produce GIS professionals who are ready to take over the responsibilities of a GIS job. This survey can help GIS managers define what skills they should look for when building the staff needed successfully to implement and operate the GIS project in an organization.

The Oklahoma State University survey results (Wikle, 1990) from 120 GIS users in private business, government agencies, and educational institutions identified six courses that more than one-half of the users perceived to be important for a GIS background. Most of the respondents identified courses in

- *Computer cartography.* More than 80 percent of GIS users identified computer cartography as an important skill to possess for a GIS background. This skill provides students with an ability to analyze and portray earth phenomena with computer software using digitizing, interactive editing, and spatial analysis skills.
- *Database management.* Almost 75 percent of GIS users identified database management skills as important for a GIS background. Since GIS involves an extensive amount of data manipulation, users felt that skills in modeling data structures and processing large volumes of data were important for their staff to possess.
- *Map reading.* More than two-thirds of the GIS users identified map-reading skills as important for a GIS background. Map projections, referencing systems, scales, and symbologies were deemed important mapping concepts to possess for GIS professionals.

More than half of the respondents also identified the following courses as being important:

- *Statistics.* About 59 percent of the respondents identified skills in error handling and confidence limits as important.
- *Aerial photograph interpretation.* About 58 percent of the respondents felt that skills in compiling information directly from source materials such as aerial photographs as being important for object recognition.
- *Computer programming.* More than half of the respondents indicated that the ability to write or modify specific applications and to troubleshoot

problems with vendor software are important skills for their GIS personnel to possess.

Other courses that were deemed important by more the a third of the GIS users included

- Remote sensing
- Physical geology
- Image processing
- Natural resources management
- Manual cartography

Managing the GIS Staff

The Kraemer et al. (1989) research identifies three important attributes concerning the structure of the GIS staff that affect the success of information systems in local government. These attributes, in-house staff versus outside consultants, functional work assignments versus task-oriented work assignments, and the rate of turnover of staff in the organization, affect the GIS professionals in the same manner and have a critical influence on the success of the ongoing operations of the GIS unit in the organization.

In-House Staff versus Outside Consultants. The lack of existing (in-house) personnel who have the skills needed to implement and operate the GIS in an organization can lead the GIS manager to the private sector to provide outside consultants for many of the roles needed in the GIS staff. Generally, this has been effective during the feasibility, cost/benefit, and vendor analysis stages of the project, primarily because outside consultants are considered to be more knowledgeable on recent technological and methodological developments concerning GIS implementation. During the needs analysis phase of the project, however, it is important to retain the knowledge learned from the detailed surveys, interviews, and analyses of the internal functional units of the organization. This knowledge, how the organization functions and what its specific information needs are, can be valuable information that should be retained over the multiyear period of planning, design, implementation, operation, and expansion of the system.

In general, in-house staffs are considered to be more responsive to the needs of the users, primarily because of their knowledge of the local culture and their dependence on the continued success of the organization. Consultants, on the other hand, are considered to be more knowledgeable on recent technological developments and methodologies that have been successful in other jurisdictions. The Kraemer et al. research, however, found that local governments that relied on in-house expertise showed greater success with computing than those who relied primarily on external technical support.

Existing GIS staff will resent the use of outside consultants when the projects that the consultants are assigned are considered to be more exciting than those assigned to the in-house staff. What in-house staff fear most is that they will be assigned the mundane tasks of maintaining existing systems (which they probably know the most about) while the consultants will be assigned the new and exciting applications of interest to top management. GIS managers must be aware of these concerns and make project work assignments accordingly balanced between in-house staff and outside consultants.

User versus Task Work Assignments. Some organizations assign staff to the needs of specific user departments while others assign staff to specific technologies regardless of the department involved. That is, a GIS manager may choose to assign the GIS staff to satisfy the needs of individual departmental users, such as the police department, the planning department, the water services department, and the sewer department, while other GIS managers may choose to assign GIS staff to specific technical tasks needed to support all users, such as database management, network management, microcomputing, land records, and geographic base files.

It should be noted that technical professionals such as GIS professionals gain more job satisfaction from assisting specific users than in developing long-range capabilities in projects that are destined to last months. As significant as some of the long-term projects are (such as land base conversion), the ability to solve particular user problems on a daily basis can distract the GIS human resources from some of the important long-range tasks required for continued expansion of the system. Consider, for example, in the role of the GIS professional who has been assigned a long-term project critical to the future of the system, but who also has been assigned the responsibility to assure that a specific user department can successfully use the system. When the telephone rings in response to a specific user-related problem with an existing application, this worker must choose between his or her attention to a long-range project (whose benefits may not be realized for many months) and the immediate needs of a user who cannot perform the system functions already implemented. Human nature favors the immediate call for help over the long-range project. The solution of the user's problem results in a satisfying environment for the GIS professional: The user is very appreciative, and normally expresses this appreciation to the GIS professional. Meanwhile, as more and more calls for help distract his or her attention from the long-range project, user managers become frustrated with the lack of progress on their major project.

This situation presents a significant dilemma to the GIS manager: how to balance the staff time required to support the daily needs of the GIS users with the long-term, major improvements needed to develop GIS capabilities for an enterprise-wide system. The task-oriented approach, assigning staff to specific technological capabilities rather than specific user departments, is commonly used in small organizations with relatively small staffs. This approach, however, is normally less effective because staff familiarity with user requirements is a vital component for developing useful applications. In addition, if one person becomes the technical expert for a specific technology (microcomputing, for example), then user contention for that expertise can cause problems in responding to problems.

Staff Turnover. Once a qualified GIS staff has been assembled, turnover becomes a critical issue for the manager. It also becomes a critical concern for the continued success of the system, since the early success stories of GIS projects have all been characterized by long-term continuity of staff in the project. Retaining staff is a result of effective leadership on the part of the GIS manager.

While some staff turnover is inevitable, primarily because of stiff competition from the private sector, there are some factors that a GIS manager can influence to assure job satisfaction by staff within the unit:

- Clearly defined career paths and management opportunities
- Staff autonomy over user relations and task completion
- Involvement with current technologies that provide technical challenges
- An understanding by staff of their role in the overall mission of the unit and of the organization as a whole

These factors define a work environment generally attractive to technically oriented people; however, each individual reacts differently to his or her work environment, and it is up to the manager to know the staff and the individual motivating forces each staff member reacts to best. There are team players and there are loners; there are intimidating personalities and there are coalition-building personalities; there are profession-devoted people and there are people who prefer more balance between their personal and their professional lives; and there are those who perform best in a highly structured workplace and others who perform best with little structure. The GIS manager must know each staff member in terms of these factors and exhibit effective leadership to provide a work environment in which they all become empowered.

This empowerment of technical people can be achieved through what Gerald Weinberg (1986) calls "organic leadership":

> Instead of leading people, as in the threat/reward model, organic leadership leads the process. Leading people requires that they relinquish control over their lives. Leading the process is responsive to people, giving them choices and leaving them in control. They are empowered in much the same way a gardener empowers seeds—not by forcing them to grow, but by tapping the power that lies dormant within them.
>
> Leadership in the seed sense is creative and productive through other people. It is an organic definition, because it works through creating an environment rather than confining itself to a few focused actions—threats or rewards—in a few specific instances to create a few specific results. (p. 12)

Managing the Work

The initial funding and work plan of the project covers a number of years to get to the point where the system can be implemented, the base map converted to digital form, and the initial applications used by operations personnel. This funding and work plan provides structure to get to the point where the "project" becomes a "system" with a service unit (the GIS staff) performing the implementation activities described in the earlier chapters of this book. The continued operation, maintenance, and management of the system in the future, however, requires a different emphasis. No longer is there a specific goal called "implementation" where all resources are being applied to activities designed for one end. Now resources are applied to a number of different activities to provide support and expansion of the system over a multiyear time period.

The process can be likened to that of purchasing a new car. Once the "implementation" activities are completed—deciding which features are needed, which model is desired, what the cost will be, which dealer will supply it, when it will be delivered, etc.—the "operations" activities begin: keeping the tank filled with fuel, scheduling maintenance work, renewing the license, adding new equipment to expand its use such as installing a cellular phone, a trailer hitch, a sun roof, etc. Then, original equipment must be replaced as tires wear out, the muffler rusts, and things do not work the way they did when new. The information processed and the activities undertaken during that "implementation" phase of purchasing a car have a singular focus: to obtain a new car. The activities associated with the "operation" of the vehicle are more diverse, somewhat repetitive, and cover a much longer time period. The "work," if you will, is different, just as the work associated with operating a mature geographic information system is different once the system has been implemented.

Managing the work required during the operational phase of a GIS consists of defining projects, developing a work plan to apply resources to the projects and schedule their completion, and preparing a budget to ensure that the financial and human resources will be available to complete the projects. While the projects may change from year to year, the process is generally the same with these three activities closely related. The budget dictates the amount of resources available to work on the projects; the work plan helps determine what resources are needed in the budget and when the projects will be completed; the projects define what work is needed to support and expand the system for the users.

Defining Projects

A project can be defined as a work activity that requires GIS resources to produce an end product of value to the user. In most cases, projects are requested by the user, but there are some projects that must be completed without a request from the user (such as routine database backups, installation of new versions of vendor software). Projects usually have a short duration (hours, days, weeks, and months) with a specific beginning and a specific ending. In order to manage the resources of the GIS unit appropriately, it is important to define all work in terms of projects. This allows management to analyze workload balance among staff members, and it also helps in determining when staff are available for new projects.

Projects can be grouped into two general categories: direct and indirect. *Direct projects* have clear beginning and ending time frames, often for a particular client, or involve a very specific activity that results in one particular product. Examples of direct projects include

- New applications requested by a user
- Changes or modifications either requested by a user or deemed necessary by technical staff
- Installation of new or replacement equipment
- Installation of new versions of software that are made available by the vendors

Indirect projects usually are less clearly defined in terms of beginning and ending time frames, often lasting an entire work-year, may be performed for all users as opposed to one particular user, or can be very repetitive and of such small duration that the work should be grouped into a larger, more general category. An example of the latter is work called "problem resolution," which can occur when the user calls a GIS staff member to fix a problem. If the problem can be fixed in less than an hour or so, it may be prudent not to define the activity as a single project but to group all such activities under the general topic of "problem resolution." If it is necessary to analyze workload of the staff in terms of which users are having the most problems, it may be necessary to define a problem resolution project for each department or office of users. Examples of indirect projects are

- Problem resolution
- Administration (usually for management and clerical staff not directly working on a specific project)
- Training (general development training as opposed to training on a specific hardware or software product that can be associated with a direct project)
- Support (usually technical activities that are very repetitive, such as performing daily database backups, preventative maintenance on equipment)

One of the most important uses of defining work in terms of projects is to have a record of the time spent by the staff by type of activity. This can be valuable information for the manager to analyze workload assignments, evaluate employee performance, and provide historical data for estimating costs and time of new projects as input to the budget and work plan preparation processes. It can also be used as a method for determining charges to users for GIS services if the organization uses a charge-back system to recover its GIS costs. To the staff, defining work in terms of projects provides a structure that is useful for communicating what is expected of them in their role as a staff member.

Projects can be used as informal "contracts" between the employee and the manager and between the GIS unit and the user (for user-requested projects). When this is the case, it is important to record pertinent information about the project that can be reviewed by manager, worker, and user alike to facilitate communication and understanding. Since a typical work-year may involve hundreds of these projects, it is necessary to establish a standard methodology for recording project information such as a "project definition form," as shown in Figure 7.4.

The project definition form contains a brief, yet comprehensive, description of the project:

- *Project title.* A short title that describes the work
- *Project number.* A unique code that can be used in an automated project management system or later memos and communications
- *Requester.* The identification of the individual or department requesting the project
- *Scope.* A paragraph that describes what the resulting product will be and how it will be used
- *Estimates.* Time, cost, and other resources that are estimated to be required to complete the project
- *Recommendation.* A statement by the GIS staff concerning the technical feasibility, or other pertinent professional observation that can useful to management in determining the priority of the project prior to committing resources to its completion
- *Signatures.* Manager, worker, and user agreement that the information about the project is acceptable to all

For direct projects requested by the user, the title, requester, and scope information should be completed by the requester while the remaining information should be completed by the GIS staff. This formal record, then, can be the source of information for developing the overall work plan for the GIS unit and the budget necessary to continue its operations.

The Annual Work Plan

After all the anticipated work for the forthcoming year has been identified and recorded onto project definition forms, it is necessary to determine the total resources required, gain budget approval to fund the work, and then schedule the work in accordance with the available resources. This is usually done in two stages: budgeting and workload planning.

Budgeting, described in more detail in the next section, starts with the

Project # _____

Date: _____

Project Title: _____

Requester: _____ Phone #: _____

Department: _____

Division/Office: _____

Type of Project: __ New Application
                       __ New Software Product
                       __ New Equipment
                       __ Change to Existing Application
                       __ Replacement of Existing Equipment
                       __ Other: _____

Project Scope: _____

_____

_____

_____

Expected Benefits: _____

_____

_____

Estimates: GIS Staff Hours _____   Cost _____

        Equipment   _____   Cost _____

                       _____   _____

                       _____   _____

        Software _____   Cost _____

        Other Costs _____   Cost _____

                  TOTAL ESTIMATED COST: _____

Recommendation: _____

FIGURE 7.4 Sample GIS project definition form.

analysis of the individual projects. The total estimated cost of each project is weighed against the potential benefits to be realized upon completion of the project. In an allocated budgeting process where each user department pays for its own projects, the user management is the first to determine if the project is worth undertaking. This is usually done early in the annual budget process, after the estimated costs for each project are determined by the GIS project office. The completed project definition form is transmitted to the requester for review and inclusion in the overall department budget. Upon review of the estimated costs, however, the department head may feel that the benefits of the project will not justify the costs and may choose to omit the project from the department budget. In a centralized budgeting process where all GIS projects are funded out of a single organization-wide budget, the decision to include or exclude a project from the budget rests with a single person or group such as the GIS manager, the IS manager, the steering committee, some administrative department head, or the top elected official. In this case, all project definition forms are reviewed and compared together as one package and rankings or priorities assigned to each project. Since there are always more projects requested than the budget can fund, this single point of decision-making will weigh not only the expected costs and benefits of each project, but will also compare the different projects to determine which are the most important for the agency as a whole, often through discussion and negotiation with the requesting department head. The priority scheme for ranking applications that was presented in Chapter 5 for developing the implementation work plan can also be used for the ongoing annual work plan as well.

Whether the allocated method for budgeting or the centralized method is used, the result of analyzing the costs and potential benefits of each project is an approval to fund the projects out of the agency's annual budget. Given this approval, then, the GIS manager must begin to schedule the work on the projects in accordance with the availability of the required staff skills needed to complete each project. This can be accomplished through the annual GIS work plan. Simply put, the annual GIS work plan is a list of all projects that were requested and approved in the budget for the next year. It contains information about each project important for effectively managing the successful completion of all projects:

- An estimate of how many staff hours will be required to complete the project
- The names of the specific GIS staff members who will be assigned the work on the project
- An estimate of the total costs of the project, including staff time, software, equipment, and other associated costs
- An estimate of when the work on the project will begin and when it is scheduled to be completed

The work plan, however, is as much a process as it is a product because it should be institutionalized within the administrative structure of the organization so that it can have widespread use throughout the agency. An annual GIS work plan can have both immediate and long-term benefits to the following groups:

- *The GIS staff.* Having a work plan for the year will provide GIS staff members with a clear understanding of the amount and type of work that is expected of them. This is especially effective when individual performance reviews are conducted to determine merit or promotional improvements. The work plan also provides staff members with an understanding of what work other staff members are assigned and how their role fits into the overall work of the office.

- *The GIS manager.* The annual GIS work plan is a tool for the GIS manager to use as a means to ensure that workload is appropriately balanced among the GIS staff; it helps the manager schedule work and determine the required staffing levels for the office.
- *User departments.* The work plan provides managers of user departments with information about which projects have been approved (in a centralized budgeting process) and when they can expect them to be completed. It is also an effective mechanism for communicating to all users details of projects other users will be receiving, in the event that they may also benefit from them.
- *Top management.* The annual GIS work plan provides top management with information that can be useful for determining the value of the system and individual projects when making decisions or setting priorities. Since the work plan summarizes all the work of the GIS office, top managers can obtain a better understanding of the need for GIS staff-level requirements.

To summarize, the annual GIS work plan is an effective management tool for determining GIS resource requirements, scheduling the work of the GIS staff, balancing workload among the various skills of the GIS staff, and communicating to the staff, the users, and top management about the work of the office and the benefits of the system. When combined with a project management system, once the work on the projects begins, it can also be effective for managing the work and the resources on an ongoing basis.

GIS Project Management

A project management system monitors the progress and reports the status of all projects included in the work plan once the work begins. It contains the same information as that recorded in the work plan, with additional information added periodically throughout the year as the work plan is being executed. This additional information on each project should include: total staff hours used to date, total costs to date, and a record of its status or problems incurred that could affect the completion date. With this additional information added on a periodic basis, the project management system provides continuous information about the status of all projects assigned to the GIS office and the workload of the staff. This is beneficial for a number of reasons:

1. It provides a reporting mechanism to user departments and top management on the current status of all approved projects.
2. It provides an accountability of the staff time spent on GIS-related activities.

3. It provides the capability to assess the impact of changes to the work plan caused by delays in completing a project, emergency requests not included in the annual work plan, extended absences of staff members, budget cutbacks, or other such unforeseen events.
4. It provides a means to identify easily the accomplishments and performance of individual staff members and the GIS staff as a whole.
5. It documents all implemented systems and applications for future reference.
6. It can be useful during the next budget cycle to assist in making estimates for similar projects and identifying uncompleted projects that will require resources in the next budget year for completion.

The project management system is thus another management and communication tool that can be of benefit for a number of people in the organization. It can be purchased as a software product from a number of different vendors, or it can be developed in house as a simple file on a personal computer. For smaller agencies with small staffs and less complex systems, crude project management systems can be established in paper form. What is important for it to be effective, however, is to ensure that the information is recorded on a regular basis (at the end of the day or week is ideal) with the assigned staff members recording their time and costs incurred by project, and that the information in the system is distributed to those who can benefit from it on a regular basis. A sample report from a typical project management system is shown in Figure 7.5.

As Of: Jan 31, 1994

Proj Number	Project Title	Requesting Office	Assigned To	Est. Cost	Est. Hrs.	Hr. to Date	Est. Compl.	Status
1001	Merge Tax/ Special Assmt Files	Finance	Myler	$12,000	250	0	Mar	To begin in Feb.
1002	Add Contract # to Permit Sys	Public Works	Torznik	$ 1,500	75	12	Feb	Need info from CCS
1009	Edge Match Quarter Sections	Engineers	Myler	$61,000	3000	87	Jun	On sched
1011	Update/Print from One Screen	Public Works	Myler	$ 2,500	12	12	Dec	Completed
1020	Install Workstation	Building Insp	Torznik	$41,800	90	10	Feb	Need quote: Vendor
...	...	...	...	...	...	...	..	...
...	...	...	...	...	...	...	..	...

FIGURE 7.5 Example of a project status report.

Budgeting for GIS Operation and Maintenance

As mentioned earlier, there are two general methods used to budget for GIS services: the allocated method, which establishes funds directly in user department budgets, and the centralized method, which establishes funds in a separate budget account for the entire organization. The method that is used can have a significant impact on how GIS resources are managed in the organization.

Centralized budgeting for GIS resources generally puts the GIS manager in control, allowing for some degree of flexibility in the technical sophistication of the system. It also can allow for some research and development work that may not otherwise be funded by any one user department. This is especially important with GIS technology because some applications of new technology may not be operational in other jurisdictions and some degree of experimentation may be necessary. As a result, centralized budgeting for GIS services can lead to larger budgets and staff than the allocated budgeting method. It can also produce an environment where the GIS manager makes most of the decisions on the priorities of implementing new applications, which can sometimes lead to conflict and confrontation between the GIS manager and user department managers.

The allocated method of budgeting for GIS services provides consistency between the amount of services used by departments and the costs of those services. This method is often called the "charge-back" method because departments are charged for the services they use, thus limiting the operational discretion of the GIS manager who then must act only as directed and financed by other departments. As a result, the allocated method of budgeting usually does not provide for any research and development activities (unless specifically requested and funded by a user); however, it often results in smaller staff sizes and budgets and can often lead to better quality in the applications developed. (When users must pay for GIS services out of their own budget, they seem to give more attention to the work, which often results in better quality.)

Whether the centralized or allocated budget method is used, there are seven major cost categories that must be funded when developing and managing the budget for GIS services:

- Labor
- Materials and supplies
- Equipment
- Software
- Maintenance of hardware and software
- Internal services
- Other expenditures

Each of these cost categories can have a "direct" component (that is, a cost directly associated with a specific project requested by a user) as well as an "indirect" component (a general cost that cannot be attributed to any project or user). Each type of component must be estimated in the budget process and eventually charged to the appropriate category in order equitably to distribute the cost of GIS services throughout the organization.

Labor

Labor costs include all costs associated with the use of human resources to perform work. Some managers, however, may chose to separate the cost of consultants and contractors from their own internal staff costs by identifying two separate categories such as "personnel" and "services." The labor category should include the costs associated with benefits in addition to the salaries of in-house staff.

Direct labor costs are those associated with the time spent by contractors or in-house staff on a particular project. This requires GIS staff, and contractors as well, to report their time by project number on a regular basis (either as part of the agency's payroll system, on the contractor's invoice, or some other means). In an allocated budgeting system, these costs are then billed to the user department—a centralized budget would only record these costs by project number for management review.

Indirect labor costs are those that are associated with time spent by the GIS staff on work that is not directly related to an approved project. As described earlier, this work may include: problem resolution, administration, training, or general system support such as preventative maintenance and daily backups. In an allocated budgeting system, these costs should be charged to indirect projects and periodically distributed to direct projects so that users are allocated these costs. An equitable allocation method for these indirect charges is to base the allocated amount on the percentage of direct hours charged to each project.

Materials and Supplies

This category includes the costs of office supplies, periodicals, books, maps, and so on to support the office of the GIS staff. Generally, these costs are all indirect costs that cannot be associated with any one project and can therefore be distributed periodically to the direct projects based upon the percentage of direct labor hours billed to the project. This provides an equitable allocation of the indirect costs since the GIS staff uses the supplies and materials in their work on the projects. Wherever possible, however, any material and supply costs that can be associated with a specific project should be charged to that project directly so that other users are not allocated that amount. (Such might be the case, for example, when a user project requires a significant amount of plotting paper, externally supplied maps, or other material.)

Equipment

All equipment purchase or lease costs, including computer hardware as well as furniture, copiers, fax machines, and so on fall into this cost category. Any equipment obtained specifically for a project (whether for use by the requesting department or use by the central GIS staff exclusively for supporting the department) should be charged directly to that department's project. Indirect equipment costs—the cost of new or replacement equipment for the centralized GIS installation—should be distributed fairly to all users. There are three methods for fairly allocating these indirect equipment costs associated with the central installation:

- *CPU utilization.* The percentage of time each department used the CPU can be obtained from the system and applied to all central equipment costs. This allocates more of the cost of the central equipment to those who use the system more, but can only be used in a centralized configuration.
- *Disk space used.* The percentage of disk space used to store each department's data can also be used to allocate the costs of the central equipment, based upon the idea that those who store more data on the system should pay more for the shared equipment. This, however, requires each database to have a single "owner"—one user who is willing to pay for the data storage—which can be a problem with enterprise-wide databases.
- *Terminals used.* Perhaps the easiest method for allocating centralized equipment costs is to base each department's share on the percentage of terminals that have to access the system. The percent of total terminals that each department has is a fair representation of the amount of use they make of the common equipment.

Software

The cost of purchased or leased software packages should be allocated in the same manner as the costs of equipment. The cost of a software package for a specific project should be assigned to that project and then charged to the department requesting the project; indirect costs of software for the central system that benefits all users (such as communications software, database management software) can be allocated using the same basis as that used for the indirect costs of centralized equipment: percent CPU utilization, percent disk space used, or percent of terminals used.

Hardware and Software Maintenance

The annual costs of maintenance on GIS hardware and software can be as much as 8 to 12 percent of the purchase price and should be allocated to users in the same manner as the costs of equipment and software: directly to the department that uses the product, or by allocating the costs among all users for centralized products.

Internal Services

Depending upon the policies and accounting practices of agencies, the costs of services performed by other departments for the GIS office may be billable to the GIS operation. Such services might include: office occupancy (including an office space charge, heating, electricity) insurance, photocopying services, printing, telephone, postage, and other similar services. These are normally general administrative costs that cannot be associated with any single project or user department. Since they usually relate to the activities of the GIS staff rather than the use of the system, it is appropriate to distribute these costs as indirect charges based upon the percent of total direct labor hours allocation methodology described under labor costs.

Other Expenses

Other expenses to be included in a GIS operations and maintenance annual budget include: travel, car allowance, training, and professional dues. Sometimes these miscellaneous expenses can be associated directly with a particular project or user department such as when a particular training program is conducted by a vendor for a particular application or when travel is required to observe a specific product that may be of interest to a user. In these cases it is possible to allocate these expenses directly to the project so that other users who may not benefit from the activity or product are not required to contribute to its cost. Generally, however, many of these miscellaneous expenses cannot be associated with a project and therefore must be distributed across all users as indirect charges based upon the percent of total direct labor hours allocation methodology described under labor costs.

Budgeting for the continual operation and maintenance of a mature geographic information system, then, requires an equitable distribution of costs to the users of the system. This can be done by allocating all the costs of the GIS operation to projects defined as either requested by users or as determined necessary to all users to support their continued use of the system. Thus the costs of operating a mature GIS operation as identified by the seven cost categories listed should be allocated either directly to projects requested by users or indirectly to the users based upon a fair allocation method that reflects actual system or other GIS resource utilization.

Summary

There are four key elements required to manage a mature geographic information system successfully after the initial implementation activities are complete:

1. Determining the most effective organizational placement of the GIS operation that can provide a continual support function for the users.

This organizational placement may differ from that used initially to design and implement the system and should reflect the management style and organizational culture of the agency.

2. Obtaining and retaining qualified and productive human resources to staff the GIS operation. This requires budgeting for the appropriate number of staff positions, producing clearly defined job descriptions, providing career progression opportunities, allowing autonomy over work assignments, and providing state-of-the-art technology for the GIS professionals.

3. Using a project management system that can identify and describe all the work of the GIS unit. This system consists of a method for defining all work into short-term, product-oriented projects that can either be associated with specific user requests or ongoing user support.

4. Preparing an annual budget that provides resources for the continual support and expansion of the system. The costs associated with this support and expansion should be equitably distributed across all users of the system, based upon their request for GIS services and their utilization of the system.

NOTES

1. The ESRI study (1989) shows that the staff size for a GIS program in governmental agencies ranges between five and nine professionals, but when individual job titles were considered, only four job titles were common among the programs: manager, senior processor, database administrator, and GIS analyst. Table 7.2 shows the percentage of programs ("sites") having staff for nine roles identified as important for a GIS program, and these four roles appeared in almost three-quarters or more of the programs. One might explain the low percentage of programmer and system administrator (53 and 51, respectfully) as being influenced by small, possibly single-station, microcomputer-based systems that do not have the technological and application complexity of much larger systems. Similarly, the low percentages of digitizer and cartographer (63 and 26, respectively) are most likely caused by programs that have these roles distributed among user departments rather than internally to the GIS office.

A surprising finding of the ESRI survey (1989) was the extent of positions with these roles that do not have GIS-related job descriptions associated with them. The second column of Table 7.2 shows the percentage of filled positions having GIS-related job descriptions. The highest of these percentages was 66 for the senior processor, and the lowest percentage, 33, was for the manager of the program. While both roles had by far the most common agreement in terms of need (90 percent, from the first column), there appears to be less agreement on their official job titles. This leads one to conclude that agencies with GIS programs have been either unwilling or unable to recognize GIS-related skills as unique or different from traditional IS skills (Huxhold, 1991).

2. While no formal degree in geographic information systems is currently being offered, several attempts to distinguish GIS-related skills apart from other, more general fields of study are being made. Most recently (in 1993), Salem State College in Salem, Massachusetts, has begun to offer an M.S. degree in "Geo Information Sciences," and

the University of Wisconsin, Milwaukee, offers a "Certificate in Urban Geographic Information Systems," administered by its Department of Urban Planning. Other academic institutions offer GIS as an emphasis option within a particular discipline (geography, geology, planning). Still others offer technical courses in GIS technology within extension or professional development programs.

REFERENCES

Byrd, C., and A. Hale (1991). Employment classification for GIS professionals: A model for implementation in state government. Unpublished paper.

ESRI (1989). *ARC News,* Summer.

Forrest, D. (1993). Developing a national human resources strategy for geomatics. *GIS World,* January.

Huxhold, W. (1991). The GIS profession: Titles, pay qualifications. *GeoInfo Systems,* March.

Izzo, J. E. (1987). *The Embattled Fortress: Strategies for Restoring Information Systems Productivity.* Jossey-Bass, San Francisco.

Kraemer, K., J. King, D. Dunkle, and J. Lane (1989). *Managing Information Systems: Change and Control in Organizational Computing.* Jossey-Bass, San Francisco.

Morgan, J., and B. Fleury (1992). Academic GIS directory. *GeoInfo Systems,* May.

Ventura, S., W. Huxhold, P. Brown, and D. Moyer (1992). Organizational issues. In *Multipurpose Land Information Systems: The Guidebook,* P. Brown and D. Moyer (eds.), chap. 8. Federal Geodetic Control Committee, North American Atmospheric and Oceanic Administration, Rockville, Md.

Weinberg, G. (1986). *Becoming a Technical Leader: An Organic Problem-Solving Approach.* Dorset House, New York.

Wikle, T. (1990). Survey targets GIS educational needs. *GIS World,* June–July.

Appendix

Additional Case Studies and Examples

Government Mapping Agencies in Canada

Government mapping agencies in Canada in the late 1970s recognized the potential savings for map production through the use of computer-aided drafting technology. At about the same time, Canadian agencies were transforming mapping from Imperial measure to metric. Additionally, new photogrammetric mapping technology capable of producing digital output was becoming available.

These changes, and the resultant demands on government mapping agencies, lead to justification and purchase of CAD technology by several provincial and Government of Canada mapping agencies. New digital mapping standards were developed individually and collectively through national coordinating committees. The private mapping companies in Canada, which provide much of the government map data, consequently also invested in new digital mapping technology.

However, the focus of the activities was to automate existing mapping processes with the intent of producing conventional (paper and film) map products and digital versions of those for future modification by those agencies or others. Initially, the end users of mapping products were for the most part not involved in the computerization process, and to a large extent were not interested in or knowledgeable about digital mapping issues.

The result was that digital mapping standards optimized map production, not geographic data management or analysis. The map data were organized as best suited drafting and map maintenance. As GIS rose in popularity and began to infiltrate resource and land management disciplines, the demand for access to the digital data and related technology increased. Many users were surprised and frustrated that the map data and associated technology were not well suited to geographic analyses and could not easily be converted to one that was suited (Burrough, 1986, pp. 13–33). In response, many provincial resource and land departments established their own systems and data to satisfy their needs. In other cases, multiagency committees and coordination projects were established to deal with the issues of map data collection and sharing and technology standards.

However, these new requirements were foreign to most mapping agencies. The business of collecting and delivering geographic data to end users changed dramatically, and new concepts, processes, and organizational structures were required to deal with the changes. What began as a relatively straightforward application of computer technology to automate portions of the map production process evolved into a process for supporting a diverse group of end users wanting to use the digital map data to build land and resource databases to conduct complex geographic analyses. In the minds of many end users the benefits were not realized. The need for the transition from the one technology to the other was not adequately foreseen or planned for.

Strategic Planning: Building a Municipal GIS, the City of Scarborough, Ontario

In March 1989, the City of Scarborough, Ontario approved a $15.8 million (Canadian) program to develop a GIS. At the same time, the City Council endorsed a mission statement for the Geo-based Information Systems (GIS) Project stating that "the City of Scarborough will develop the best user-oriented and user-defined municipal GIS in North America." The "best" could mean the most frugal in a financial sense, the best response time or the easiest to use. Scarborough aims, however, to measure the system based on how closely the GIS functionality relates to the business needs of the municipality and its agencies and clients. City Council also appointed a GIS project manager with 15 years of public works experience.

Scarborough's GIS is intended not only to serve all departments of the municipality, but will embrace the needs of a variety of external agencies, other levels of government, and the private sector. All departments of the municipality and many of these external agencies are participating in a variety of joint application development sessions to put the system in place. Further, they have created a number of agreements that will result in cost sharing of the enterprise approach to GIS.

Within the municipality, the GIS is being built as part of an overall information systems technology strategy. In other words, the GIS is not a separate function, but rather a set of geographic databases and applications within the context of an information systems program. Prior to initiating the project, the total information requirements of the municipality were identified. Four geographic database structures were identified: topographic, cadastral, structures, and engineering. An additional nine databases will meet the need for nongeographic information, including such things as finance, payroll, and human resources.

The strategic issues identified and addressed in modeling and project management of the Scarborough GIS were

- *Project management*. While there are a multiplicity of projects and individuals, related projects will use a matrix organization structure. Effective project management will ensure a level of communication to all participants.
- *Commitment*. Since the implementation will require long-term commitment of financial and human resources, paramount importance was given to the need for frequent communication of tangible results and for a participatory approach.
- *Hardware and software*. Since there is no single set of hardware or GIS software to meet the needs fully, it is strategic that the GIS team and the users maintain a broad knowledge of existing facilities to ensure the responsiveness to the needs.
- *Training*. Concerns centered around the development and retention of skilled GIS staff, as well as a wide range of potential users. Training encompassed attitudinal issues, realistic expectations, and an understanding of the commitment and energies required.
- *Agreements with others*. Success is dependent upon a high level of participation to ensure the achievement of objectives at a reduced cost. Agreements need to be carefully documented and endorsed by all concerned.
- *Security*. Providing security and confidential access to the related databases was determined to be a strategic issue. The question of liability regarding the provision of property-related information will be addressed.
- *Integration/flexibility*. The ability to integrate existing computerized and manual systems into the GIS model was important. Sufficient flexibility in the design to provide for a high level of support to meet changing needs was deemed strategic.
- *Users' needs*. This issue centered around the definition for accuracy, user-defined systems, ownership, responsibility for updating, and shared data.

GIS Guidelines for Assessors

Prior to a two-use study to prepare GIS guidelines for assessors, the International Association of Assessing Officials developed the following list of questions in September 1989 (International Association of Assessing Officials, 1989):

Administration:
1. How does a GIS benefit the assessor?
2. What changes will be required in the way the assessment office operates to make the best use of a GIS?
3. Will GIS help me carry out my responsibilities more effectively?
4. Will it reduce the paper flow?

5. What are some limitations of GIS?
6. Who are the beneficiaries of GIS?
7. How does a jurisdiction determine if a GIS is an appropriate investment?
8. What are some key factors which affect the cost of implementing a GIS?
9. Do the potential benefits of a GIS really outweigh the development costs?
10. How can a small-to-medium jurisdiction justify the purchase of a GIS?
11. How do I get support for developing a GIS?
12. How do I finance a GIS?
13. Can we sell some of the data we manage to outside customers?
14. How do you price information in a GIS?
15. How do I coordinate a GIS project?
16. Who should be involved in developing a GIS?
17. Who is in charge of defining, documenting, and enforcing standards?
18. What are some key factors which affect a time schedule in implementing a GIS?
19. Should I have a specific GIS manager during the contracting phase?
20. If I invest in GIS hardware/software now, how long will it last?
21. Can I simply add enhancements to the system, or will I be faced with scrapping it and buying all new technology?
22. Can a GIS be installed as a turnkey system, or must it be customized to meet the unique requirements of the jurisdiction?

Training and education:
23. How will my staff adjust to this new technology?
24. How can I assure current employees that GIS will not eliminate their jobs?
25. What are the education requirements for staff and management?
26. What levels of training are required for GIS?
27. How much time must be devoted to training?
28. Can my present staff be trained to operate the GIS?
29. What reference materials are available to help educate me in GIS?
30. How do I deal with inconsistent terminology?

Hardware and software:
31. How do I know which hardware and software to buy?
32. Can I run the software on my existing in-house hardware?
33. How adaptable are current systems? Can they be upgraded as technology changes?
34. Can microcomputers be used for smaller-scale GIS projects?
35. Software vendors are now splitting products into different components. How do I determine which components I need for my project?

36. Since the price of hardware and software is falling, why not wait until next year to purchase a system?

Database design:
37. What does a GIS database contain?
38. What is the networking of databases?
39. How is a GIS database created?
40. What are GIS database standards?
41. What approaches are available for automating property maps?
42. How do I evaluate the quality of existing base maps?
43. Will the GIS incorporate the existing parcel number in my jurisdiction?
44. How do I resolve property boundary discrepancies in developing an automated property map?
45. How accurate does the land base have to be?
46. How does the cost vary for different levels of accuracy?
47. How can existing data be converted to digital form, and what are some of the restrictions?
48. How do I link my existing data files to a GIS?
49. How difficult will it be to interface our existing appraisal software with GIS?
50. How does a GIS relate to my CAMA (computer-assisted mass appraisal) system?
51. Will the GIS and CAMA be redundant?
52. Can I transfer data back and forth between systems? Is this process practical?
53. Must I use the same system for mapping and data manipulation?

Use and selection of consultants:
54. Do I need a consultant?
55. Should I use the same consultant/vendor to select and install the GIS?
56. What terminology do I need to know to converse intelligently with vendors/consultants?
57. How do I know that I am conveying my needs?
58. How do I identify the best consultant/vendor for my needs?
59. Is a utility company consultant/vendor the same as a consultant for a municipality?
60. How do I determine if the consultant/vendor is reputable?

Vendor selection:
61. Who should digitize the property maps?
62. Should the job be done in house, or should an outside contractor be considered?
63. What are the advantages/disadvantages of contracting out the mapping and data entry (data conversion) compared with doing it in house?

System maintenance:
  64. What commitment must be made towards maintaining the system after it has been implemented?
  65. Who should support the GIS? Should it be my staff, the data processing staff, a vendor, or a consultant?
  66. What is the cost of maintaining a GIS database?

Cooperative efforts:
  67. Can the cost of a GIS be shared with other agencies?
  68. How do I find other agencies that are interested in participating in a GIS?
  69. How can the rest of the community/jurisdiction benefit from a parcel-based GIS?
  70. If working in a multiuser GIS, how do we decide which data elements to store?
  71. Should we try to eliminate duplicate data (in a multiuser GIS)?
  72. How does one determine maintenance responsibilities (in a multiuser GIS)?
  73. What kinds of computerized files are available from other agencies?
  74. How do I make use of digital data collected by other organizations?
  75. Is it true that data sharing between different organizations is no longer a problem because different GIS packages can pass data back and forth by using "interchange formats"?

Examples of Procurement

Centra Gas, Winnipeg, Manitoba

Some organizations have taken this basic process as outlined previously and developed some unconventional, but very effective, approaches to procurement. Centra Gas, Winnipeg, Manitoba, developed an innovative approach for communicating their needs and gathering information about suppliers. They invited suppliers of products and services, one at a time, to their offices in Winnipeg, where they set up a room displaying the organization's processes and products that would need to be supported by the GIS and related information technology products. A company representative spent the morning presenting the process and products and answering questions. In the afternoon, the suppliers of the day then presented their products and services. At the conclusion of the meeting, interested suppliers were given guidelines for preparing a proposal document. With more than 20 companies involved, this process required better than a month to complete, but given the scope of the procurement, it was probably the most efficient way to communicate that much information. The process eliminated the need to prepare a comprehensive and complex procurement document, since respondents had reviewed the Centra Gas business process and requirements first hand. The replies were better formulated and specific to the needs of Centra Gas. Note that if the company's procurement policy had forbidden any face-to-face communication with suppliers prior to making a selection, this process would not have been possible.

Most organizations will not undertake a process like the one at Centra Gas. More common will be a process during which written documents are exchanged between suppliers and the purchaser.

Manitoba Ministry of Forests' Example of a Requirement Statement and RFP Table of Contents

An example of written documents exchanged between the parties of a procurement using a set of specifications and an RFP outline developed by the Manitoba Ministry of Forests for procurement of a GIS system is presented here (Lamont, 1988).

Input information list:
1. Forest resource inventory maps
2. Forest wildfire maps
3. Natural disaster maps (wind, drought, flood)
4. Forest insect and disease maps
5. Silviculture activity maps
6. Timber management activity maps
7. Soils maps
8. Land and resource use maps
9. Forest resource development proposal maps
10. Utility corridor maps (proposed and/or actual)
11. Forest capability Canada land inventory map
12. Satellite imagery (Landsat and variations)

User requirement list outline:
1. Map file preparation
   - Location description
   - UTM coordinate value entry
   - All stands and types
   - Stand attribute identifiers
   - Photo center and numbers, section covers, and road allowances
   - Data to be stored in polygons, attribute files or overlays
   - Edge matching
2. Area calculation of all polygons including single line features such as railroads
3. Map production
   - Plotted on Mylar
   - Plotted on paper
   - All established boundaries

- All labels
- All location identifiers

4. Data manipulation
   - Analysis of effect of management decisions
   - Consider analysis
   - Area analysis
   - Buffer zone analysis
   - Wildfire loss analysis
   - Overlay analysis (125,000 polygons at a time)

5. Updating
   - Review map and associated attribute files
   - Opportunity for future satellite digital data application
   - Review status and ownership boundaries
   - Enter new features as they are constructed
   - Wildfire area and loss calculation/revision of original forest cover data

6. Thematic mapping
   - Color code and shading code options to identify selected forest stand attributes
   - Determine a new code or shading code based on results of polygon overlays

7. Site-specific information
   - Integration of town street maps into a database where specific items at large scale (1:500–1:2,000) are to be positioned

8. Data entry options
   - Keyboard data entry without tieup of graphic workstation terminal
   - From tape files

9. System functions
   - Digitizing—point and line features
   - Polygonization—areas connected to form polygons, built-in editing routines for overshoots, undershoots, and open polygons
   - Drawing/editing—commands to support including editing verification
   - Alphanumeric data entry: stand attribute data entered through keyboard on tape file
   - Editing: duplicate lines/polygons, open polygons, overshoots, under-shoots
   - Symbolizing
   - Merging and dropping line
   - Weeding/thinning
   - Subdivision of polygons
   - Browsing
   - Check pointing and restart
   - Generation of graphic entities
   - Lines from digitized point
   - Draw circles from center point

- Define corridors along straight or convoluted lines
- Complex polygons
- Search by attribute
- Redefine attribute
- Centroid positioning
- Windowing
- Scale changes—full range performed in the computer
- Projections—to increase graphic capabilities for large-area mapping
- Measurements
- Perimeter and area
- Balance areas to a predetermined total
- Measuring and storing straight line distance between digitized points
- Polygon overlay processing
- Combine different overlays for analysis of data and plotting results of combined analysis—grid analysis
- Corridor analyzing
- Edge matching
- Rubbersheeting
- Updating
- Display and plotting
- User-defined calculations relating to digitized information and attributes

10. Reports required
    - Area listing by stand numbers
    - Summaries of area
    - Area and volume of stands
    - Summaries of volume
11. Equipment requirements
    - Digitizing workstations
    - Data entry workstations
    - Disk storage capacities
    - Tape drive
    - Query terminal
    - Printer—speed
    - Plotter—speed, color pens
12. Software
    - operating system software
    - Specialized application software
    - Third-party software to be included

Proposal preparation: The proposal should address all the concerns and fulfill all the requirements specified by the RFP. The proposal should contain the following information:

1. Introduction
2. Business plan
   - Vendor's background
   - Approach to GIS system
3. Response to user requirement
   - Hardware configuration
   - System software
   - Application software
   - User system function requirements
   - Data storage
   - Growth potential
   - Communication
   - Security and backup
   - Vendor services: references, installation, training, maintenance, documentation
   - Implementation schedule
   - Financial considerations
   - Contractual considerations
   - Programming requirements
   - Alternatives
   - Other additional information that the vendor feels is pertinent to the proposal
4. Description of systems hardware, software, and functional capabilities
5. Product description
6. Brochures and spec sheets
7. Management overview

Three copies of the proposal must be submitted to Manitoba Data Services. Proposals should be placed in binders with tabs separating the major sections.

Technology Selection Criteria by the Alberta Planning Board

Technology and Vender Selection Process: Request for Information

The Alberta Planning Board LRIS pilot project team translated the conceptual system design into technical specifications to be sent as a *request for information* to known vendors of land information technology and related services. The request for information was to contain a description of the pilot project, the procedures to be tested, and the details of the content of the proposal document.

A selection committee consisting of experienced users and technology experts was formed to select the technology and vendor. The selection committee devised a set of criteria, which they ranked using the analytical hierarchical process. This was the same process as was used to choose the pilot project area and to analyze the planning topics. The selection committee discussed and agreed upon the criteria, their descriptions, and their weights. These creatures are presented with their assigned weights in Table A.1.

Among the criteria was *participation*—the commitment and resources within the scope of the proposal. The selection committee was seeking a commitment to seeing the proposed strategy work, and to seeing the project through to a successful conclusion as well as an ability to provide the human resources necessary to implement the strategy. The vendor was assumed to take on a partnership role with the other project participants. Since the evaluation of alternatives during the project may require some additional effort, it was considered essential that the vendor be an active participant in the project.

As project coordinator, the Alberta Research Council requested vendors that proposals be submitted describing a total hardware and software combination that would be capable of handling both short- and long-term system requirements, within the unique needs and objectives of the pilot project. The approach taken with the request for information was to provide the vendors with as much flexibility as possible to configure a deliverable system that would have the capability to complete the pilot project successfully, to grow to meet the needs of the participants in the future, and to be transferable to other municipalities throughout Alberta. The reason was that the vendors know their product and corporate commitment to this type of project, so no attempt was made to restrict the imaginativeness of the resulting system or approach.

To assist in the evaluation process, the proposal was to be divided into two sections: a technical proposal; and the corporate contribution and approach to involvement.

Table A.1   The Selection Criteria and Weights for Selecting Technology and Vendor

Criterion	Weight (%)		
Strategy	73.5		
Transferability			36.6
Ease of use			10.7
Maintenance			8.4
Lifespan			6.3
Versatility			3.5
Phasing			3.2
Growth			2.7
Price			1.8
Technology		24.4	
The firm		6.4	
Participation		6.1	
Contribution	20.7		
Cost*	5.8		
	100.0	73.5	36.6

* Cost was considered to the extent that it had to fall within an acceptable price range dictated by the project budget. Principal concern was value for dollar instead of lowest price.

The Technical Proposal. Within the technical proposal, the vendors described the software, its capabilities and how it would satisfy the application requirements outlined. The vendors presented the hardware configurations as described, including

- Individual configurations for each of the three proposed sites
- Communications strategy between each of the proposed sites
- Communications with government computers within the LRIS network

One of the primary objectives of the pilot project was to test the transferability of the effort, technology, data, and skill sets required to facilitate the implementation of useful LRIS technology from the pilot project to other municipalities within the province. With this in mind, the affordability of the proposed hardware, and the hardware already in place within the pilot agencies, had to be reconciled with what would be affordable if this type of work was undertaken by other jurisdictions.

REFERENCES

Burrough, P. A. (1986). *Principles of Geographic Information Systems for Land Resources Assessment.* Oxford University Press, Oxford.

International Association of Assessing Officers (1989). GIS Guidelines Questions, internal working document, September.

Lamont, R. H. (1991). Avoiding surprises in selecting and setting up a GIS. In *GIS Applications in Natural Resources,* Michael Heit and Art Shortreid (eds.), pp. 71–74. GIS World, Fort Collins, Colo.

INDEX

Administrative placement of a GIS, 205–6
Aerial-photography services, procurement of, 180
Alberta Planning Board, 201, 239–40
Alberta Research Council, 240
AM/FM systems, 34
   procurement of, 181
Annual workplan, 218–21
Antenucci, J., 134
Applications, 144–54
   documentation of, 147–49
   implementation of, 188–97
   priorities of, 148–49
   selection of, 105–7
Applications software specifications, 193
Applied development services, procurement of, 180
Aronoff, S., 201
Arthur Young & Company, 125, 127
Automated mapping software, 150
Automated procedures, 34–35, 45, 49, 73, 229
   transition of analog to digital, 49

Benchmarking, 187
   definition of, 201
Blanchard, K., 161
Booth, G., 39–40, 41
Budgeting, 223–26
   cost categories, 223
Building permit inspections, 13–14
Bullets (planning tasks), 188
Burg, M., 200
Business area analysis, 125
Business manager, GIS, 165–66
Byrd, C., 210–11

Cadastral features, 143, 144
Cadastral mapping services, procurement of, 180–81
CAD. See Computer-aided drafting
CADD. See Computer-aided drafting and design

CAM. See Computer-aided mapping
Campbell, H., 29–30
Cartographic technician, role of, 172
CASE. See Computer-assisted software engineering
Caston, A., 23
Centra Gas, 234–35
Change management, 54, 58
   participative vs. directive change, 161
   reducing behavioral problems, 162
Change orders, 116–17
Clerical and administrative support staff, role of, 172–73
COGO. See Coordinate geometry
COM. See Computer output microfilm
Commitment vs. compliance, 99–100
Communication, 100, 197
Communications software, 151
Competitive performance, 38
Computer graphics systems, 35
Computer output microfilm (COM), 153
Computer performance, gains in, 10–11
Computer system administrator, role of, 169–70
Computer technology, evolution of stages in, 59–60
Computer-aided mapping (CAM), 34, 50
Computer-aided drafting (CAD), 34, 49–50, 229
Computer-aided drafting and design (CADD), 12, 34, 49
   procurement of, 181–82
Computer-assisted software engineering (CASE), 123–24
Conceptual overview, 91
   elements of, 104
   issues, 104
   techniques to facilitate, 103
Conceptual planning, 52
Conferences, benefits of, 103
Consultants
   procurement of, 177–80
   use of, 101, 103, 213–14

Contour lines, 143
Control surveys, procurement of, 181
Coordinate geometry (COGO), 123, 150, 154
Cost–benefit analysis, 69, 70
Costs
    data conversions, 134
    equipment, 224–25
    hardware, 224–25
    internal services, 226
    labor, 224
    maintenance, 225
    materials and supplies, 224
    software, 225
Crime statistics, 16–17
Critical success factors, 104, 118–19
Croswell, P., 148–49
Current physical model, 127, 128–31
    definition of, 129
Customized programming software, 151
Customizing software, 195

Data compilation and conversion, 191–92
Data compilation and maintenance, reducing
    cost and efforts, 191–92
Data conversion, 87
    costs, 134
    services, procurement of, 179–80
Data inputs, 146
Data management software, 150
Data management strategy, 9–10
Data models, 7–10
    attributes of, 8
    common data platform, 8, 10
    creation of, 14
    logical, 8, 9
    municipal government, 26
    perception structures, 7, 13–14
    spatial entities, 8
    system design, 21
Data needs, determining, 134–37
Data outputs, 146–47
Data planning, within the strategic plan, 52
Data standards, 8–9
Data structure, 124
Database, definition of, 41
Database administrator, role of, 169–70
Database loading, 194–95
Database management systems (DBMS), 12,
    34
Database structure, 8, 23
DBMS. See Database management system
de Bono, E., 52
Dealing with the unknown, 113–16
Decision support systems (DSS), 73

Definitions
    benchmarking, 201
    current physical model, 129
    database, 41
    feasibility, 84
    geographic information system, 3, 5
    geography, 23
    geomatics, 200
    information system, 41
    management framework, 95
    multiparticipant projects, 74–75
    paradigm, 6, 84
    participants, 95
    project and systems management, 216–17
    situational analysis, 54, 57
    systems design, 121
Delegation, reasons against, 96–97
DeMarco T., 101, 117
Design specifications, 87
Dickinson, B., 199
Dickson, P., 162
Digitizers, 153–54
Direct projects, 216
Document imaging, 12
    procurement of, 183
Documentation standards, 93
Driving forces, 48
Drucker, P., 38
DSS. See Decision support systems

EDI. See Network communications and
    electronic data interchange
Engineering surveys, procurement of, 181
Environmental Systems Research Institute
    (ESRI), 208–9, 227
Errors, cost of, 112–13
ESRI. See Environmental Systems Research
    Institute

Facilities management. See Outsourcing
Feasibility, 55, 67–74
    definition of, 84
    financial, 68, 69–71
    institutional, 72–74
    measuring, 68–74
    technical, 71–72
Functional analysis, 127–28
Functional business model, 127–28
Functional decomposition, 132–33
Functional needs, determining, 144–46
Functions requiring maps, 129

General design, 92–93
    specifications of, 159

Geocoding, 7
Geographic information system (GIS)
  analyst, role of, 171–72
  benefits of , 3–4, 44, 46, 50, 53, 70
  champion, role of, 98–99
  concepts of, 6–7
  conceptual model of, 92
  data management, 7–10
  definitions of, 3, 5
  effect of, on roles in the organization, 54
  elements of, 5
  as enabling technology, 11
  evolution of, 34–35
  external factors, 63
  ground rules for successful design, 39
  hardware, procurement of, 181
  implementation of, 5
    failure of, 45
    issues, 11–14, 30
  job descriptions, 208–13
  maintaining support, 106
  market estimates, 44
  needs inventory, 134–37
  as new way of doing business, 51
  organizational strategy, 11–14
  paradigm, 5, 6, 21, 33
  paradigm shift, 50, 51, 199
  placement of, in organization, 204–7
  procurement of, 181–83
  returns on investment, 54
  technologies, 10–11
Geography, definition of, 23
Geomatics, definition of, 200
Geomatics Industry Association of Canada
    (GIAC), 211–12
Geoprocessing, 12
Georeferencing, 7
GIS. See Geographic information system
Global Positional System (GPS), 11, 12 23
  procurement of, 182
Gotlieb, C., 84
GPS. See Global Positioning System

Hale, A., 210–11
Hardware and software, procurement and
    installation of, 87
Hardware performance specifications, 193
Hardware requirements, 151–54
Hersey, P., 161
Hesjedal, O., 54
Human Resources Department
  getting GIS positions, 211
  involvement, 162
  training planning, 176

Huxhold, W., 61, 62, 63, 126, 134, 210
Hypsography, 143

IAAO. See International Association of
    Assessing Officials
Image analysis, procurement of, 182
Implementation failure, 120
Implementation framework, 91, 241
Implementation management framework, 162
Implementation plans/planning
  detailed design, 190
  documentation, 108
  example, 109
  importance of, 107
  issues and steps, 88–89
  process, 90
  purpose of, 88
  review points, 112
  task definitions, 113
  transition from planning to construction,
    160–63
Implementation roles and responsibilities,
    95–102
Implementation schedule, 89
Implementation tasks (bullets), 188
IMPS. See Integrated municipal information
    system
Indirect projects, 216–17
Information engineering, 124–26
  objectives of, 125
  phases of, 125–26
Information flow, 135–36
Information management, 33
Information needs inventory, 134–37
Information needs matrix, 139–40
Information oriented design, 34
Information Resources Management (IRM),
    33–36
Information system
  definition of, 41
  planning, 125
Information technology (IT), 33
  benefiting business procedures, 53
  changes to, 64
  planning, 35–36
  stages of maturity, 59–60
Installation/procurement, hardware/software,
    87
Integrated municipal information systems
    (IMPS), 26
Interface with existing packages software,
    151
International Association of Assessing
    Officials (IAAO), 31–33, 40–41, 121,

International Association of Assessing
    Officials (*continued*), 231–34
IRM. *See* Information Resources Manage-
    ment

IT. *See* Information technology
Izzo, J., 207–8

JAD. *See* Joint Application Design
Job descriptions, 208–13
Job titles, 227–28
Joint Application Design (JAD), 101–2, 118
Joint ventures, 83

Keen, P., 88
Kramer, K., 60–64, 207

Lamont, R., 235
Lead agency in multiparticipant project, 79
Leadership, 215–16
Legal surveys, procurement of, 181
Lind, A., 118
Lister, T., 101, 117
Loading the database, 194–95
Logical model
    current, 127, 132–33
    new, 127, 133–34

Management
    decision points, 159
    middle, role of, 49
    middle, training needs, 175–76
    senior, role of, 97–98
    senior, training needs, 175
Management audit, 57
Management consulting, 177–79
Management framework, 31, 39 52–56
    changes to, 160
    definition of, 95
    example of, 97
    implementation of, 162
Management information systems (MIS), 33,
    34–35
Management philosophy, 57
Management practice, 37–40
Managing GIS staff, 213–15
Managing uncertainty, 113–16
Manitoba Ministry of Forests, 235–38
Map inventory, 138–39
Map overlays, 6
Maps, functions requiring, 129
Martin, J. , 35–36, 37, 41, 73, 124
Massachusetts, 29
McClure, C., 35–36, 124

Menu design software, 151
Milwaukee , 41, 208–9
MIS. *See* Management information systems
Mix state. *See* States of computing
    technology
Models/modeling. *See also* Data models
    current logical model, 127, 132–33
    current physical model, 127, 133–34
    functional business model, 127–34
    new logical model, 127, 133–34
    process, 92
    real world, 8
Monuments, 141
Motivating, 200
Multimedia, 12
    procurement of, 183
Multiparticipant projects, 54, 63, 72
    characteristics of, 76
    committee structure, 79–80
    coordination vs. joint venture, 83–84
    definition of, 74–75
    issues, 46
    lead agency, 79
    motivation for, 79
    organizational structure, 81
    planning for, 80–82
    scope of benefits, 54
    strategic planning for, 74–84
    temporary vs. permanent, 82–83
Municipal government physical components,
    14–18

National Geodetic Reference System
    (NGRS), 141
Needs analysis, 87, 89
Network communications and electronic data
    interchange (EDI), 12
NGRS. *See* National Geodetic Reference
    System
Nolan, R., 59–60

Oklahoma State University, 212
Onsrud, H., 28
Ontario, 200
    example of outsourcing, 163
Operating system characteristics, 193
Operational audit, 127–28
Operational organizational placement of a
    GIS, 205
Organizational behavior, 37
Organizational culture, 57
Organizational planning, within the strategic
    plan, 52
Organizational structure, 81

Organizational theory (OT), 36–37
OT. *See* Organizational theory
Outsourcing, 85, 163, 200
    when to use, 163, 213–14

Paradigm, definition of, 6, 84. *See also*
    Geographic information system (GIS),
    paradigm
Participants, definitions of (sponsors, end
    users/clients, project management
    system designers, and implementors data
    providers), 95
PERT. *See* Project evaluation and review
    technique
Photogrammetric mapping services,
    procurement of, 181
Pilot projects, 93, 116–17, 196, 239–40
Pinto, J., 28
Placement of a GIS, 205–7
Planimetric features, 141–43
Planning, tools for, 123–24
Planning horizon, 68
Planning tasks (triggers), 188
Plotters, 153
POLARIS. *See* Province of Ontario Land
    Registration Information System
Policy framework, 31
Position descriptions, 227–28
Private sector, physical components, 18–21
Process consultation, 84
Process flow, 123
Process-oriented design, 34
Procurement/installation
    aerial-photography services, 180
    AM/FM systems, 180
    applied development services, 180
    cadastral features, 180–81
    case examples, 235–38
    computer-aided drafting and design, 182
    consultants, 177–80
    control surveys, 181
    data conversion services, 179–80
    document imaging, 183
    engineering surveys, 181
    global positioning systems, 182
    hardware, 87, 181
    image analysis, 182
    legal surveys, 181
    multi-media, 183
    photogrammetric mapping services, 181
    policy, 183–86
    process, 184–88
    project mapping services, 181
    quality assurance and quality control, 179

raster scanning, 183
software, 87
supervisory control and data acquisition,
    182–83
technology and services, 176–88, 192–94
Productivity, 38
Programmers, role of, 170–71
Project and systems management
    benefits of, 221–22
    definition of, 216–17
    procurement of, 179
    project management system, 221–22
    software, 108, 110–11
    tips, 188–89
Project evaluation and review technique
    (PERT), 41
Project management, 39
Project manager, project leader, role of,
    166–67
Project mapping services, procurement of,
    181
Project reporting, 197
Project scope, 67–74
Project teams, 99–102
    issues, 101
Property statistics, 16–17
Prototypes, 196
Province of Ontario Land Registration
    Information System (POLARIS), 200
Public Land Survey System (PLSS), 143

Quality assurance and quality control
    procurement of, 179

Raster scanning, procurement of, 183
Reducing behavioral problems, 162
Remote sensing, 12
Request for bids (RFB), 184
Request for information (RFI), 239
Requests for proposal (RFP), 184, 185–87
    example of, 234–38
Resource allocation, 112
RFB. *See* Request for bids
RFI. *See* Request for information
RFP. *See* Requests for proposal
Risk, technical, 115–16
Risk management, 113–16
Roles and responsibilities, 95–102
    GIS champion, 98–99
    GIS implementation, 164–74
    project teams, 99–102
    senior management, 95–102

SCADA. *See* Supervisory control and data
     acquisition
Scarborough (city), 55–56, 65, 230–31
Schein, E., 84
Scott, W. R., 119
Seminars, 103
Service state. *See* States of computing
     technology
Shared land base, 140–41
Simmons, J., 162
Situational analysis, 54, 57–58, 81, 88
     applying, 65
     definitions of, 54, 57
Skill state. *See* States of computing
     technology
Skunkworks, 101, 118
Software and hardware needs, software,
     142–49
Software customization, 195
Software engineering, 123
Software requirements, 150–51
South Carolina, 208–9
Spatial analysis software, 151
Sponsors, definition of, 95
Spot elevation, 143
Staffing
     balancing assignments, 214–15
     economizing on smaller projects, 173
     growth rate/demand, 211
     job satisfaction, 214
     needs, 164–74
     roles, 209–10
     skills match, 100
     turnover, 215–16
Standards development, 189–90
Start-up plan, 91–92
States of computing technology
     mix state, 60, 62
     service state, 60, 61
     skill state, 60–61
Strategic approach, 54, 81
Strategic placement of a GIS, 206–7
Strategic planning, 47, 52–56, 81, 89–91
     benefits of, 52
     for multiparticipant projects, 74–84
     review points, 157–59
Strategic state. *See* States of computing
     technology
Strategic vision, 55, 65–67, 88
Structured analysis, 123
Structured design, 123
Structured programming, 123
Supervisory control and data acquisition
     (SCADA), 12

procurement of, 182–83
Survey and mapping services, procurement
     of, 181
Survey control features, 141
SYMAP. *See* SYnagraphic MApping system
Symbology creation software, 151
SYnagraphic MApping system (SYMAP), 7
System testing, 195–96
System users, role of, 173
Systems architecture, 93–94
Systems design
     benefits of, 121
     definition of, 121
     detailed, 94
     elements of, 121–22
     ground rules of, 39–40, 87
     obstructions to, 126
     tools, 123–24

Tapscott, D., 23
Technical consulting services, procurement
     of, 180
Technical risk, 115–16
Technology architecture, 193–94
Technology planning, within the strategic
     plan, 52
Technology testing, 194
Teranet Land Information Services, Inc., 200
Text processing, 12
Thematic mapping, 6–7
Tips on project management, 188–89
Topographic features, 143
Topology, 7
Townsend, R., 157
Training, 87, 94
     courses required, 212–13
     degrees offered, 227–28
     needs, 174–76
     role of vendors, 176
Triggers (planning tasks), 188
Tveisdal, S., 54

Uncertainty, managing, 113–16
University of California, Irvine, 60
Urban and Regional Information Systems
     Association (URISA), 31, 40–41
Urban Information Systems (URBIS), 60–64
Urban Information Systems Inter-Agency
     Committee (USAC), 26–28, 80
URBIS. *See* Urban Information Systems
URISA. *See* Urban and Regional Information
     Systems Association
USAC. *See* Urban Information Systems
     Inter-Agency Committee

User interface, 93

Vermont, 29
Vision statement, development of, 65–67

Walton, R., 53

Weinberg, G., 215–16
Wikle, G., 212
Wisconsin Land Records Modernization
    Project, 78
Workstations, 152–53